Haralampos Hatzikirou

Cellular automata for the analysis of cancer invasion

Haralampos Hatzikirou

Cellular automata for the analysis of cancer invasion

Mathematics against cancer

Südwestdeutscher Verlag für Hochschulschriften

Impressum/Imprint (nur für Deutschland/ only for Germany)
Bibliografische Information der Deutschen Nationalbibliothek: Die Deutsche Nationalbibliothek verzeichnet diese Publikation in der Deutschen Nationalbibliografie; detaillierte bibliografische Daten sind im Internet über http://dnb.d-nb.de abrufbar.

Alle in diesem Buch genannten Marken und Produktnamen unterliegen warenzeichen-, marken- oder patentrechtlichem Schutz bzw. sind Warenzeichen oder eingetragene Warenzeichen der jeweiligen Inhaber. Die Wiedergabe von Marken, Produktnamen, Gebrauchsnamen, Handelsnamen, Warenbezeichnungen u.s.w. in diesem Werk berechtigt auch ohne besondere Kennzeichnung nicht zu der Annahme, dass solche Namen im Sinne der Warenzeichen- und Markenschutzgesetzgebung als frei zu betrachten wären und daher von jedermann benutzt werden dürften.

Verlag: Südwestdeutscher Verlag für Hochschulschriften Aktiengesellschaft & Co. KG
Dudweiler Landstr. 99, 66123 Saarbrücken, Deutschland
Telefon +49 681 37 20 271-1, Telefax +49 681 37 20 271-0
Email: info@svh-verlag.de
Zugl.: Dresden, TU, Diss., 2009

Herstellung in Deutschland:
Schaltungsdienst Lange o.H.G., Berlin
Books on Demand GmbH, Norderstedt
Reha GmbH, Saarbrücken
Amazon Distribution GmbH, Leipzig
ISBN: 978-3-8381-1769-0

Imprint (only for USA, GB)
Bibliographic information published by the Deutsche Nationalbibliothek: The Deutsche Nationalbibliothek lists this publication in the Deutsche Nationalbibliografie; detailed bibliographic data are available in the Internet at http://dnb.d-nb.de.

Any brand names and product names mentioned in this book are subject to trademark, brand or patent protection and are trademarks or registered trademarks of their respective holders. The use of brand names, product names, common names, trade names, product descriptions etc. even without a particular marking in this works is in no way to be construed to mean that such names may be regarded as unrestricted in respect of trademark and brand protection legislation and could thus be used by anyone.

Publisher: Südwestdeutscher Verlag für Hochschulschriften Aktiengesellschaft & Co. KG
Dudweiler Landstr. 99, 66123 Saarbrücken, Germany
Phone +49 681 37 20 271-1, Fax +49 681 37 20 271-0
Email: info@svh-verlag.de

Printed in the U.S.A.
Printed in the U.K. by (see last page)
ISBN: 978-3-8381-1769-0

Copyright © 2010 by the author and Südwestdeutscher Verlag für Hochschulschriften Aktiengesellschaft & Co. KG and licensors
All rights reserved. Saarbrücken 2010

Lattice-gas cellular automata models for the analysis of cancer invasion

DISSERTATION

zur Erlangung des akademischen Grades

Doctor rerum naturalium
(Dr. rer. nat.)

vorgelegt

der Fakultät Mathematik und Naturwissenschaften
der Technischen Universität Dresden

von

MSc.-Technomathematiker **Haralambos HATZIKIROU**

geboren am 6. März 1978 in Thessaloniki, Hellas

Gutachter: Prof. Dietmar Ferger
Prof. Dieter Wolf-Gladrow

Eingereicht am: 25/03/2009

Tag der Verteidigung: 10/07/2009

Acknowledgments

I would like to express my gratitude first to Prof. Andreas Deutsch for the advices, support and freedom I have enjoyed during these years that allowed me to follow my interests and points of view. I also would like to thank him for trying to put some order in my innate tendency towards chaos.

I also would like to extend my gratitude to Arnaud Chauviere for all the infinite discussions and the development of new ideas that have contributed so much to this thesis - ok...he could have been a better play maker but he was the only one available. Moreover, i would like to thank David Basanta for his contribution to the maturation of my ideas concerning tumor evolution and for the thousand coffees that we have drunk together - hopefully he will give out the secret of his diet and i will loose some weight. Special thanks to Fernando Peruani for sharing with me his interesting ideas and his idealistic perspective about science - i hope he is going to learn not to foot on sea urchins. I would like also to thank Sebastiano De Franciscis for working together and his critical view - and also express my support for this poor Napolitan "alien" in Germany. Finally, special thanks to Marco Tektonidis for his valuable collaboration and for his friendship - and i wish him to stop changing apartments each two months.

I would like to thank my medical collaborators Prof. Carlo Schaller and MD Dr. Matthias Simon for introducing me to tumor biology and to the "fabulous" clinical world. In particular, it was a great - and at the same time shocking - experience to attend a brain tumor surgery.

Particularly, i would like to express my gratefulness to Prof. Dieter Wolf-Gladrow for reviewing my thesis and for his valuable comments. Moreover, i would like to thank Prof. Dietmar Ferger and Dr. Anja Voss-Boehme for accepting reviewing this thesis. I thank also Kathrin Padberg for reviewing and correcting my thesis.

I would like to express my thanks to Gustavo Sibona, Luca Barberis, Simona Faini and Subrata Nandi for their collaboration and the time that we spent together. To Joern Starruss, Tobias Klauss, Michael Kuecken, Lutz Brusch, Joachim Peter, Christian Hoffmann, Carsten Mende, Peggy Thiemt, Ronny Zschitzschmann, Perla Zerial, Niloy Ganguli, Michael Wurzel and Edward Flach for their helpfulness and creating such a pleasant atmosphere.

Last but not least, i would to thank my family for their support and love all these years. I would like to dedicate this thesis to them. Specially, i would like to thank them for teaching me how to stand directly after falling and finding always the bright side of life. Finally, i would like to finish with my life's motto:

Carpe Diem

Contents

1 Introduction   **1**
1.1 The problem: cancer invasion 1
1.2 The questions 3
1.3 The existing models 3
1.4 The approach 5
1.5 Overview of the thesis 6

I Lattice-gas cellular automata   **9**

2 Lattice-gas cellular automata: Basics   **11**
2.1 Introduction 11
2.2 Lattice-gas cellular automata 12
    2.2.1 States in lattice-gas cellular automata 14
    2.2.2 Dynamics in lattice-gas cellular automata ... 15
2.3 Discrete kinetic theory 17
    2.3.1 Stochastic description 18
    2.3.2 Lattice Boltzmann equations 19
2.4 Summary 20

3 Random cell motion   **21**
3.1 Introduction 21
    3.1.1 Brownian motion 22
    3.1.2 Modeling of Brownian motion 23
3.2 A LGCA model for randomly moving cells 25
    3.2.1 Microscopic dynamics 26
    3.2.2 Mesoscopic dynamics 26
3.3 Characterization of random motion 28
    3.3.1 Individual cell motion 28
    3.3.2 Collective transport 31
3.4 Summary 40

4 Growth processes   **41**
4.1 Introduction 41
4.2 A LGCA growth process 43
    4.2.1 LGCA dynamics 44
    4.2.2 Microdynamical equations 45
4.3 Mean-field analysis 46
    4.3.1 Well-stirred system 46
    4.3.2 Spatially distributed system 48

4.4	Mean-field approximations and macroscopic behavior	55
4.5	Summary	57

II Tumor invasion   59

5 The impact of environment on tumor cell migration   61
- 5.1 Introduction . . . 61
  - 5.1.1 Types of cell motion . . . 62
  - 5.1.2 Mathematical models of cell migration . . . 64
  - 5.1.3 Overview of the chapter . . . 65
- 5.2 LGCA models of cell motion in a static environment . . . 65
  - 5.2.1 Model I . . . 70
  - 5.2.2 Model II . . . 70
- 5.3 Analysis of the LGCA models . . . 74
  - 5.3.1 Model I . . . 74
  - 5.3.2 Model II . . . 78
- 5.4 Summary . . . 80

6 The impact of migration and proliferation   83
- 6.1 Introduction . . . 83
- 6.2 The LGCA model . . . 84
  - 6.2.1 LGCA dynamics . . . 85
  - 6.2.2 Micro-dynamical equations . . . 86
- 6.3 Results . . . 87
  - 6.3.1 Simulations . . . 87
  - 6.3.2 Mean-field analysis . . . 88
  - 6.3.3 Traveling tumor front analysis . . . 96
- 6.4 Summary . . . 98

7 Mechanisms of tumor invasion emergence   101
- 7.1 Introduction . . . 101
- 7.2 Why not mutations? . . . 105
- 7.3 The LGCA model . . . 108
  - 7.3.1 The microenvironment: Oxygen concentration . . . 108
  - 7.3.2 Cell dynamics . . . 108
- 7.4 Simulations . . . 111
- 7.5 Analysis . . . 115
  - 7.5.1 Mean-field approximation . . . 116
  - 7.5.2 Macroscopic dynamics . . . 116
- 7.6 Summary . . . 117
  - 7.6.1 Biological evidence . . . 118
  - 7.6.2 Therapy . . . 119

Contents

8 *In vivo* tumor invasion   **121**
    8.1   Introduction 121
    8.2   The algorithm 123
         8.2.1   Image segmentation 124
         8.2.2   Tumor modeling 127
         8.2.3   Shape analysis 130
         8.2.4   Evolutionary algorithm 131
    8.3   Results 134
    8.4   Summary 136

9 Discussion   **139**
    9.1   Summary of the thesis 139
    9.2   Can statistical mechanics help us to understand tumors? 141
    9.3   Discussion and outlook 142
         9.3.1   Mathematical analysis of LGCA 143
         9.3.2   Tumor invasion modeling 143

A Calculation of random walk LBE   **145**

B Calculation of equilibrium distributions   **147**
    B.1   Equilibrium distribution of Model I 147
    B.2   Equilibrium distribution of Model II 149
    B.3   Analytical parameter estimation in Model II 149

C Details of tumor growth model in Chapter 6   **153**

Bibliography   **155**

CHAPTER 1

Introduction

Contents

1.1	The problem: cancer invasion	1
1.2	The questions	3
1.3	The existing models	3
1.4	The approach	5
1.5	Overview of the thesis	6

1.1 The problem: cancer invasion

Cancer describes a group of genetic and epigenetic diseases, characterized by uncontrolled growth of cells, leading to a variety of pathological consequences and frequently death. Cancer has long been recognized as an evolutionary disease [Nowell 1976]. Cancer progression can be depicted as a sequence of traits or phenotypes that cells have to acquire if a neoplasm (benign tumor) is to become an invasive and malignant cancer. A phenotype refers to any kind of observed morphology, function or behavior of a living cell. Hanahan and Weinberg [Hanahan 2000] have identified six cancer cell phenotypes: unlimited proliferative potential, environmental independence for growth, evasion of apoptosis, angiogenesis, invasion and metastasis (fig. 1.1). The temporal order of tumor cells acquiring new capabilities remains still unknown.

In this thesis, we concentrate on the *invasive phase of solid tumor growth*, i.e. the cancer phase that is related to the appearance of the invasive phenotype. Invasion is the main feature that allows a tumor to be characterized as malignant. The progression of a benign tumor and delimited growth to a tumor that is invasive and potentially metastatic is the major cause of poor clinical outcome in cancer patients, in terms of therapy and prognosis. Understanding tumor invasion could potentially lead to the design of novel therapeutical strategies. However, despite the immense amounts of funds invested in cancer research, the intracellular and extracellular dynamics that govern tumor invasiveness *in vivo* remain poorly understood.

Biomedically, invasion involves the following tumor cell *processes*:

- tumor cell proliferation,

Chapter 1. Introduction

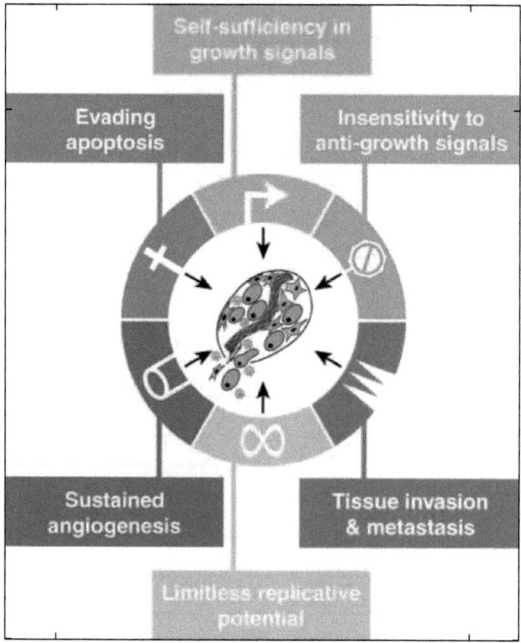

Figure 1.1: Hanahan and Weinberg have identified six possible types of cancer cell phenotypes: unlimited proliferative potential, environmental independence for growth, evasion of apoptosis, angiogenesis, invasion and metastasis (Reprinted from [Hanahan 2000], with permission from the authors.)

- tumor cell migration, which for instance is a result of down-regulation of cadherins[1] that leads to the loss of cell-cell adhesion and

- tumor cell-extracellular matrix (ECM) interactions, such as cell-ECM adhesion, and ECM degradation/remodeling, by means of proteolysis. These processes allow for the penetration of the migrating tumor cells into host tissue barriers, such as basement and interstitial stroma [Friedl 2004].

Tumor invasion facilitates the emergence of metastases, i.e. the spread of cancer cells to another part of the body, through the lymphatic and the blood vessel network, and the formation of secondary tumors. It is obvious that tumor invasion

---
[1]**Cadherins**: Important class of transmembrane proteins. They play a significant role in cell-cell adhesion, ensuring that cells within tissues are bound together. They are dependent on calcium (Ca^{2+}) ions to function; hence their name.

comprises a central aspect in cancer progression. However, invasive phenomena occur not only in pathological cases of malignant tumors but also during physiological morphogenesis and wound healing.

1.2 The questions

Tumor invasive behavior is characterized from the combined effect of tumor cell proliferation, tumor cell migration and cell-microenvironment interactions. This combination makes the understanding of tumor invasion extremely challenging. In order to analyze the tumor invasion problem we break it down to subproblems that correspond to the impact of the basic invasion processes on tumor behavior, described in subsec. 1.1. In this thesis, we focus on the following subproblems/questions of tumor invasion:

(**Q1**) *What is the impact of the environment on tumor cell migration?* In the definition of tumor invasion (subsec. 1.1) environmental cues play a significant role. Of great interest is the influence of the ECM on the migrating behavior of tumor cells [Friedl 2004].

(**Q2**) *How does tumor proliferation and migration influence tumor's invasive behavior?* Uncontrolled proliferation is the essential requirement for tumor development. Combined with tumor cell migration - no ECM effects are considered - provides the minimal prerequisites for tumor invasion. The main question that arises is how fast the tumor expands or, in other words, what is the invasion speed of a tumor.

(**Q3**) *Which are the mechanisms of tumor invasion emergence?* Typically, tumor invasion appears during the late stages of carcinogenesis. Of ultimate importance is the knowledge about mechanisms and the environmental conditions that trigger the progression from benign neoplasms to malignant invasive tumors.

(**Q4**) *How can we compute* in vivo *tumor invasion?* The prediction of *in vivo* tumor evolution is the ultimate challenge. The quest for reliable, quantitative prediction methods of clinical tumor growth concentrates the interest of numerous researchers.

The above list of subproblems provides a "modular" approach to the understanding of the tumor invasion problem. In the following, we describe our attempt to approach and to answer the above questions.

1.3 The existing models

Cancer invasion has been recognized as a complex phenomenon that involves processes which occur at different spatio-temporal scales. Mathematical modeling provides invaluable tools in the analysis of complex phenomena. Mathematical models allow for the description and the linking of these levels. Therefore, in this thesis, we use mathematical modeling as a microscope to gain insight into the complex tumor invasion dynamics.

In particular, one can distinguish *molecular*, *cellular* and *tissue* scales [Hatzikirou 2005, Preziosi 2003]:

- The molecular scale refers to phenomena at the sub-cellular level and concentrates on molecular interactions and resulting phenomena, such as alterations of signaling cascades and cell cycle control, gene mutations, etc. In tumor invasion, the down-regulation of cadherins provides an example of a molecular process.

- The cellular scale refers to cellular interactions and therefore to the dynamics of cell populations mediated by adhesion, contact inhibition, chemotaxis, haptotaxis etc.

- The tissue scale focuses on tissue level processes taking into account macroscopic quantities, such as tissue volumes, blood flow etc. Typically, macroscopic quantities are mathematically expressed in continuous terms. Continuum phenomena include cell convection and diffusion of nutrients and chemical factors, mechanical stress (e.g. edema) and the diffusion of metastases.

Tumor dynamics involve phenomena at all three scales. For example, genetic alterations may lead to invasive cells (molecular scale) that are able to migrate (cellular scale) and interact with diffusible or non-diffusible signals (tissue scale). Models that deal with phenomena at multiple scales are called multi-scaled.

Mathematical models of tumor invasion have a relatively history. However, there exists already a large literature of modeling studies focusing on this stage of cancer. Here we recall solely the models that are related to the main focus of this thesis, i.e. the questions (Q1)-(Q4):

Models related to (Q1) Recently, several authors have developed mathematical models to analyze the effect of the ECM on tumor cell motion [Perumpanani 1999, Sherratt 2001, Turner 2002]. A particularly interesting case is the influence of the brain environment (fiber tracks) on the evolution of glioma tumors, investigated by [Swanson 2002, Wurzel 2005].

Models related to (Q2) An important aspect of tumor invasion is the calculation of the invasion speed. Macroscopic models have been used to model the spatio-temporal growth of tumors study tumor's invasion speed, usually assuming that tumor invasion is a wave propagation phenomenon [Sherratt 1992, Perumpanani 1996, Marchant 2000]. These models connect the tumor invasive speed with the tumor cell proliferation and migration rates.

Models related to (Q3) The models that study the emergence of tumor invasion investigate the conditions under which tumor cells acquire the essential capabilities which allow a tumor to be characterized as invasive (subsec. 1.1).

In [Smallbone 2007] is studied the emergence invasive behavior interpreted as emergence of a glycolytic tumor cell population that is able to acidify the neighboring tissue. In this study the focus is on the emergence of the capability of tumor cells to degrade/remodel their environment. In [Basanta 2008a], the authors study the circumstances under which mutations that confer increased motility to cells can spread through a tumor composed of rapidly proliferating cells.

Models related to (Q4) Recently, several computational models have been develop in order to predict the *in vivo* tumor's spatiotemporal evolution. Swanson et al. [Swanson 2002] modeled proliferation and migration of brain tumors based on actual clinical data and calculated the glioma spatio-temporal evolution of real anatomical geometries. Lately, multi-scale approaches attempt to provide computational platforms that allow for *in vivo* predictions [Alarcón 2003, Mansury 2006, Frieboes 2007].

The above models of tumor invasion can be sorted in two coarse categories. The first category includes simplified tumor models that mathematical analysis is feasible but lack of biological relevance. The other category deals with computational models that provide high level of biological details but, due to the complexity of these models, analysis is often prohibitive. Our aim is to find a compromise between the high complexity of the biological details of tumor invasion and the mathematical abstraction. In particular, our module-oriented approach allows for the development of analyzable models related to the different aspects of tumor invasion (as presented in subsec. 1.1). The synthesis of these modules-models allows for the development of more complex, biologically relevant model that hopefully contributes to the understanding of tumor invasion. In the following, we present in detail the philosophy of our approach.

1.4 The approach

In this thesis, our approach is based on the mathematical abstraction of the tumor invasion problem. We reduce the tumor invasion problem to the *interplay* of the two main tumor cell processes, i.e. proliferation and migration. Our strategy is to describe the relevant phenomena that influence tumor cell migration and proliferation (e.g. ECM cues) at the cellular scale and analyze the emergent macroscopic invasive tumor behavior. Cellular Automata (CA) are considered mathematics provide tools that allow for a consistent transition from a cellular description to a macroscopic scale.

Cellular automata, and more generally cell-based models, provide an discrete modeling approach, where a micro-scale investigation is allowed through a stochastic description of the dynamics at the cellular level [Deutsch 2005]. In particular, CA define an appropriate modeling framework for tumor invasion since they allow for:

- CA rules can mimic the tumor processes at the cellular level. This fact allows for the modeling of an abundance of experimental data that refer to cellular and sub-cellular processes related to tumor invasion.

- The discrete nature of CA can be exploited for investigations of the boundary layer of a tumor. Bru et al. [Bru 2003] have analyzed the fractal properties of tumor surfaces (calculated by means of fractal scaling analysis), which can be compared with corresponding CA simulations to gain a better understanding of the tumor phenomenon. In addition, the discrete structure of CA facilitates the implementation of complicated environments without any of the computational problems characterizing the simulation of continuous models.

- Motion of tumor cells through heterogeneous media (e.g ECM) involves phenomena at various spatial and temporal scales [Lesne 2008]. These cannot be captured in a purely macroscopic modeling approach. On the other hand, discrete microscopic models, such as CA, can incorporate different spatio-temporal scales and they are well-suited for simulating such multiscale phenomena.

- CA are paradigms of parallelizable algorithms. This fact makes them computationally efficient.

Here, we focus on an important class of CA, the so-called lattice-gas cellular automata (LGCA), which recently have been proposed as models for the analysis of biological phenomena [Deutsch 2005]. In contrast to traditional CA, LGCA provide a straightforward and intuitive implementation of cell migration and interactions. Finally, the structure of LGCA facilitates the mathematical analysis of their behavior.

1.5 Overview of the thesis

In this thesis, the general scope is to shed light on important aspects of tumor invasion, under the magnifying glass of mathematical modeling and analysis. The main processes involved in tumor invasion, related to tumor cell migration, cell-environment interactions and tumor cell proliferation, evolve on different scales. In order to understand tumor invasion dynamics, it is important to use mathematical tools that allow for modeling sub-cellular or cellular processes and to analyze the emergent macroscopic behavior. Individual-based models, and especially CA, are well-suited for this task. In this thesis, we use a special type of CA models, the so-called lattice-gas cellular automata [Deutsch 2005], which facilitate analytical investigations allowing for deeper insight into the modeled phenomena.

The complete comprehension of tumor invasion dynamics is very difficult since tumor invasion is a complex phenomenon. Therefore, we split the tumor invasion problem into subproblems (questions) that correspond to the impact of the basic invasion processes on tumor behavior, described in subsec. 1.1, i.e.

1.5. Overview of the thesis

(Q1) What is the impact of environment on tumor cell migration?

(Q2) How do tumor proliferation and migration influence a tumor's invasive behavior?

(Q3) Which are the mechanisms of tumor invasion emergence?

(Q4) How can we compute *in vivo* tumor invasion?

The analysis of the above subproblems provides information that allows us to improve our understanding of the tumor invasion problem.

This thesis is divided in two parts. In the first part, we present the tools for the mathematical analysis of LGCA in basic biological concepts, such as random cell motion and growth. The second part of the thesis, deals with tumor invasion and it provides answers to the questions (Q1)-(Q4). In particular, this thesis consists of the following chapters:

Chapter 2 Here we introduce the nomenclature of lattice-gas cellular automata and basic elements of the discrete kinetic theory, in particular the derivation of the lattice Boltzmann equation (LBE).

Chapter 3 The scope of this chapter is two-fold: (i) to introduce a LGCA of random cell motion and (ii) to provide mathematical tools for the analysis of the LGCA model. The analysis of the random motion LGCA model is conducted by estimating the cell motility rate and by deriving its macroscopic behavior. The mathematical tools and concepts presented in this chapter serve as a basics for the understanding of the following chapters of this thesis.

Chapter 4 In this chapter, we introduce a LGCA that models a population of duplicating and randomly moving cells. We demonstrate how a mean-field approximation can yield insight into the formation of spatial patterns and calculate important macroscopic observables for the biological growth process. In particular, we address the role of the diffusion strength in the approximation by distinguishing well-stirred and spatially distributed cases.

Chapter 5 We present a LGCA as a microscopic model of cell migration together with a mathematical description of different tumor environments. We study the impact of the various tumor environments (such as extracellular matrix) on tumor cell migration by estimating the tumor cell dispersion speed for a given environment (cp. Q1).

Chapter 6 We study the effect of tumor cell proliferation and migration on the tumor's invasive behavior by developing a simplified LGCA model of tumor growth (cp. Q2). In particular, we derive the corresponding macroscopic dynamics and we calculate the tumor's invasion speed in terms of tumor cell proliferation and migration rates. Moreover, we calculate the width of the invasive zone, where the majority of mitotic activity is concentrated, and it is found to be proportional to the invasion speed.

Chapter 7 We investigate the mechanisms for the emergence of tumor invasion in the course of cancer progression (cp. Q3). We conclude that the response of a microscopic intracellular mechanism (migration/proliferation dichotomy) to oxygen shortage, i.e. hypoxia, maybe responsible for the transition from a benign (proliferative) to a malignant (invasive) tumor.

Chapter 8 We propose an evolutionary algorithm that estimates the parameters of a tumor growth LGCA model based on time-series of patient medical data (in particular Magnetic Resonance and Diffusion Tensor Imaging data). These parameters may allow to reproduce clinically relevant tumor growth scenarios for a specific patient, providing a prediction of the tumor growth at a later time stage (cp. Q4).

Chapter 9 We present a summary of the main thesis results. Moreover, we discuss the plausibility of the mathematical framework used, in this thesis, for the analysis of the tumor invasion problem. Finally, we discuss possible extensions of the questions investigated in this thesis.

Part I

Lattice-gas cellular automata

CHAPTER 2

Lattice-gas cellular automata: Basics

Contents

2.1	Introduction	11
2.2	Lattice-gas cellular automata	12
	2.2.1 States in lattice-gas cellular automata	14
	2.2.2 Dynamics in lattice-gas cellular automata	15
2.3	Discrete kinetic theory	17
	2.3.1 Stochastic description	18
	2.3.2 Lattice Boltzmann equations	19
2.4	Summary	20

2.1 Introduction

Lattice-gas automata have been introduced as a fully discrete fluid model, the so-called HPP model [Hardy 1973] (named after Hardy, de Pazzis, and Pomeau). The HPP model used a square lattice, where the gas particle motion in orthogonal directions is defined by momentum conserving collisions. The advantage of the model against traditional cellular automata was the intuitive implementation of the particle motion. HPP failed to simulate Navier-Stokes due to the inadequate symmetry of the square lattice. Subsequently, Frisch, Hasslacher, and Pomeau (FHP) developed a hexagonal LGCA that was sufficient to yield the Navier-Stokes equation in the macroscopic limit [Frisch 1987]. The strength of the lattice-gas method lies in unraveling the potential effects of movement and interaction of individuals[1] (e.g. particles, cells). In traditional cellular automaton models implementing movement of individuals is not straightforward, as one node in the lattice can typically only contain one individual, and consequently movement of individuals can cause collisions when two individuals want to move into the same empty node. In a lattice-gas model this problem is avoided by having separate channels for each direction of movement. The channels specify the direction and magnitude of movement, which

---
[1]LGCA models are individual-based models. Without loss of generality, in this chapter we are going to use the term particle to denote the modeled individuals.

may include zero velocity (resting) states. Moreover, LGCA impose an exclusion principle on channel occupation, i.e. each channel may at most host one particle.

The transition rule of a LGCA can be decomposed into two distinct steps. The first is an *interaction* step (mathematically called operator) that updates the state of each channel at each lattice node. Particles may change their velocity state (*reorientation*) or undergo a birth/death stochastic process (*reactions*) as long as they do not violate the exclusion principle. The second is the *propagation* step, particles move synchronously into the direction and by the distance specified by their velocity state. The propagation step is deterministic and conserves mass and momentum, which is in general not the case for the interaction (reorientation, reactions) operators. Synchronous transport prevents particle collisions which would violate the exclusion principle (other models have to define a collision resolution algorithm). LGCA models allow parallel synchronous movement and updating of a large number of particles.

The basic idea of lattice-gas automaton models is to mimic complex dynamical system behavior by the repeated application of simple local transport and interaction rules. The reference scale (fig. 2.1) of the LGCA models introduced in this thesis is that of a finite set of cells. The dynamics evolving at smaller scales (intracellular) are included in a coarse-grained manner, by introducing a proper stochastic interaction rule. This means that our lattice-gas automata impose a microscopic, though not truly molecular, view of the system by conducting fictive microdynamics on a regular lattice. In the following, we present in detail the formalism and the nomenclature of LGCA that is used within this thesis.

2.2 Lattice-gas cellular automata

A lattice-gas cellular automaton is a cellular automaton with a particular state space and dynamics. Therefore, we start with the introduction of cellular automata which are defined as a class of spatially and temporally discrete dynamical systems based on local interactions. In particular, a cellular automaton is a 4-tuple $(\mathcal{L}, \mathcal{E}, \mathcal{N}, \mathcal{R})$, where

- $\mathcal{L}$ is a regular lattice of nodes (discrete space),

- $\mathcal{E}$ is a finite set of states (discrete state space); each node $\mathbf{r} \in \mathcal{L}$ is assigned a state $\mathbf{s} \in \mathcal{E}$,

- $\mathcal{N}$ is a finite set of neighbors on the lattice $\mathcal{L}$ (neighborhood); Moore and von Neumann neighborhoods are typical neighborhoods on the square lattice,

- $\mathcal{R}$ is a deterministic or probabilistic map

$$\mathcal{R} : \mathcal{E}^{|\mathcal{N}|} \to \mathcal{E}$$
$$\{s_i\}_{i \in \mathcal{N}} \mapsto s,$$

2.2. Lattice-gas cellular automata

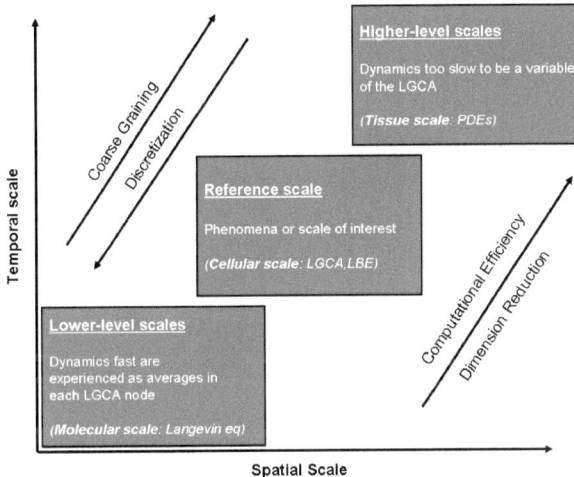

Figure 2.1: The sketch visualizes the hierarchy of relevant scales for the LGCA models introduced in this thesis (see text for details).

which assigns a new state to a node depending on the states of all its $|\mathcal{N}|$ neighbors (local rule), where the symbol $|\cdot|$ denotes the cardinality of a set.

The temporal evolution of a cellular automaton is defined by applying the function $\mathcal{R}$ synchronously to all nodes of the lattice $\mathcal{L}$ (homogeneity in space and time).

2.2.1 States in lattice-gas cellular automata

We define a d-dimensional regular lattice $\mathcal{L} = L_1 \times ... \times L_d \subset \mathbb{Z}^d$, where $L_1, ..., L_d$ are the lattice dimensions. Typically in this thesis, we will refer to two-dimensional models ($d = 2$). Particles move on the discrete lattice with discrete velocities, i.e. they hop at discrete time steps $k \in \mathbb{N}$ from a given node to a neighboring one. A set of velocity channels $(\mathbf{r}, \mathbf{c}_i)$, $i = 1, ..., b$, is associated with each node $\mathbf{r} \in \mathcal{L} \subset \mathbb{Z}^d$ of the lattice. The parameter b is the *coordination number*, i.e. the number of velocity channels on a node which coincides with the number of nearest neighbors on a given lattice. In particular, the set of velocity channels for the square lattice as considered here, is represented by the two-dimensional channel velocity vectors $\mathbf{c}_1 = \begin{pmatrix} 1 \\ 0 \end{pmatrix}$, $\mathbf{c}_2 = \begin{pmatrix} 0 \\ 1 \end{pmatrix}$, $\mathbf{c}_3 = \begin{pmatrix} -1 \\ 0 \end{pmatrix}$, $\mathbf{c}_4 = \begin{pmatrix} 0 \\ -1 \end{pmatrix}$, $\mathbf{c}_5 = \begin{pmatrix} 0 \\ 0 \end{pmatrix}$, (see fig. 2.2). In addition, a variable number $\beta \in \mathbb{N}_0 = \mathbb{N} \cup \{0\}$ of rest channels (zero-velocity channels), $(\mathbf{r}, \mathbf{c}_i)$, $b < i \leq b+\beta$, with $\mathbf{c}_i = \{0\}^\beta$ may be introduced. Furthermore, an exclusion principle is imposed. This requires, that not more than one particle can be at the same node within the same channel. As a consequence, each node $\mathbf{r}$ can host up to $\tilde{b} = b + \beta$ particles, which are distributed in different channels $(\mathbf{r}, \mathbf{c}_i)$ with at most one particle per channel. Accordingly, state $\mathbf{s}(\mathbf{r})$ is given by

$$\mathbf{s}(\mathbf{r}) = \left(\eta_1(\mathbf{r}), ..., \eta_{\tilde{b}}(\mathbf{r})\right) =: \boldsymbol{\eta}(\mathbf{r}),$$

where $\boldsymbol{\eta}(\mathbf{r})$ is called *node configuration* and the quantities $\eta_i(\mathbf{r}) \in \{0, 1\}, i = 1, ..., \tilde{b}$ are called *occupation numbers*, which are Boolean variables that indicate the presence ($\eta_i(\mathbf{r}) = 1$) or absence ($\eta_i(\mathbf{r}) = 0$) of a particle in the respective channel $(\mathbf{r}, \mathbf{c}_i)$. Therefore, the set of elementary states $\mathcal{E}$ of a single node is given by

$$\mathcal{E} = \{0, 1\}^{\tilde{b}}.$$

The *node density* is the total number of particles present at a node $\mathbf{r}$ and time $k \in \mathbb{N}$ denoted by

$$n(\mathbf{r}, k) := \sum_{i=1}^{\tilde{b}} \eta_i(\mathbf{r}, k).$$

The *global configuration* over the lattice at time k is given by

$$\boldsymbol{\eta}(k) := \{\boldsymbol{\eta}(\mathbf{r}, k)\}_{\mathbf{r} \in \mathcal{L}}.$$

2.2. Lattice-gas cellular automata

For any node $\mathbf{r} \in \mathcal{L}$, the nearest lattice neighborhood $\mathcal{N}_b(\mathbf{r})$ is a finite list of neighboring nodes and is defined as

$$\mathcal{N}_b(\mathbf{r}) := \{\mathbf{r} + \mathbf{c}_i \,:\, \mathbf{c}_i \in \mathcal{N}_b, \, i = 1, \ldots, b\} \,.$$

Fig. 2.2 gives an example of the representation of a node on a two-dimensional lattice with $b = 4$ and $\beta = 1$, i.e. $\tilde{b} = 5$.

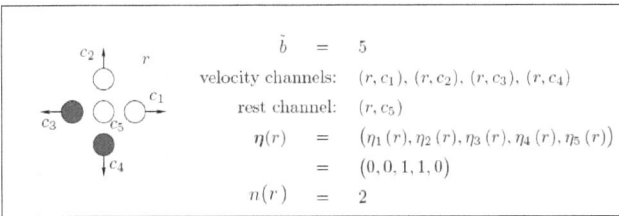

Figure 2.2: Node configuration: channels of node $\mathbf{r}$ in a two-dimensional square lattice ($b = 4$) with one rest channel ($\beta = 1$). Gray dots denote the presence of a particle in the respective channel.

In multi-component LGCA, ς different types of particles σ reside on separate lattices $\mathcal{L}_\sigma$ and the exclusion principle is applied independently to each lattice. The state is given by

$$\mathbf{s}(\mathbf{r}, k) := \boldsymbol{\eta}(\mathbf{r}, k) = \{\boldsymbol{\eta}_\sigma(\mathbf{r}, k)\}_{\sigma=1\ldots\varsigma} = \{(\eta_{\sigma,1}(\mathbf{r}, k), \ldots, \eta_{\sigma,\tilde{b}}(\mathbf{r}, k))\}_{\sigma=1\ldots\varsigma} \in \mathcal{E} = \{0, 1\}^{\tilde{b}\varsigma} \,.$$

2.2.2 Dynamics in lattice-gas cellular automata

The dynamics of a LGCA arises from repetitive application of superpositions of local (probabilistic) *interaction* and deterministic *propagation* (transport) steps applied simultaneously to all lattice nodes and at each discrete time step. The definitions of these steps have to satisfy the exclusion principle, i.e. two or more particles are not allowed to occupy the same channel.

According to a model-specific *interaction* rule ($\mathcal{R}^C$), particles can change channels (see fig. 2.3) and/or are created or destroyed. The interaction rule can either model a rule that regulates the particle motion or the birth/death of particles. The temporal evolution of a state $\mathbf{s}(\mathbf{r}, k) = \boldsymbol{\eta}(\mathbf{r}, k) \in \{0, 1\}^{\tilde{b}}$ in a LGCA is determined by the temporal evolution of the occupation numbers $\eta_i(\mathbf{r}, k)$ for each $i \in \{1, \ldots, \tilde{b}\}$ at node $\mathbf{r}$ and time k. Accordingly, the pre-interaction state $\eta_i(\mathbf{r}, k)$ is replaced by the post-interaction state $\eta_i^C(\mathbf{r}, k)$ determined by

$$\eta_i^C(\mathbf{r}, k) = \mathcal{R}_i^C(\{\boldsymbol{\eta}(\mathbf{r}, k) | \mathbf{r} \in \mathcal{N}_b(\mathbf{r})\}), \quad (2.1)$$

$$\mathcal{R}^C(\{\boldsymbol{\eta}(\mathbf{r}, k) | \mathbf{r} \in \mathcal{N}_b(\mathbf{r})\}) = \left(\mathcal{R}_i^C(\{\boldsymbol{\eta}(\mathbf{r}, k) | \mathbf{r} \in \mathcal{N}_b(\mathbf{r})\})\right)_{i=1}^{\tilde{b}} = \mathbf{z},$$

realized with probability $\mathbb{P}(\{\boldsymbol{\eta}(\mathbf{r},k)|\mathbf{r}\in\mathcal{N}_b(\mathbf{r})\}\to\mathbf{z})$ and $\mathbf{z}\in(0,1)^{\tilde{b}}$, which is the time-independent transition probability of the pre-interaction node state to the post-interaction one.

Figure 2.3: Example of a possible interaction of particles at a two-dimensional square lattice node $\mathbf{r}$; filled dots denote the presence of a particle in the respective channel. No confusion should arise by the arrows indicating channel directions.

In the deterministic *propagation* or streaming step (P), all particles are moved simultaneously to nodes in the direction of their velocity, i.e. a particle residing in channel $(\mathbf{r},\mathbf{c}_i)$ at time k is moved to another channel $(\mathbf{r}+m\mathbf{c}_i,\mathbf{c}_i)$ during one time step (fig. 2.4). Here, $m\in\mathbb{N}_0$ determines the *single particle speed* and $m\mathbf{c}_i$ the *translocation* of the particle. Because all particles residing at the same velocity channels move the same number m of lattice units, the exclusion principle is maintained. Particles occupying rest channels do not move since they have "zero velocity". In terms of occupation numbers, the state of channel $(\mathbf{r}+m\mathbf{c}_i,\mathbf{c}_i)$ after propagation is given by

$$\eta_i(\mathbf{r}+m\mathbf{c}_i,k+\tau)=\eta_i^{\mathrm{P}}(\mathbf{r},k),\quad \mathbf{c}_i\in\mathcal{N}_b(\mathbf{r}), \qquad (2.2)$$

where $\tau\in\mathbb{N}$ is the automaton's time-step. We note that the propagation operator is mass and momentum conserving. Hence, if only the propagation step was applied then particles would simply move along straight lines in directions corresponding to particle velocities.

Combining interactive dynamics (C) (2.1) with propagation (P) (2.2) implies that

$$\eta_i(\mathbf{r}+m\mathbf{c}_i,k+\tau)=\eta_i^{\mathrm{CP}}(\mathbf{r},k). \qquad (2.3)$$

This can be rewritten as the *microdynamical difference equations*

$$\eta_i(\mathbf{r}+m\mathbf{c}_i,k+\tau)-\eta_i(\mathbf{r},k)=\eta_i^{\mathrm{CP}}(\mathbf{r},k)-\eta_i(\mathbf{r},k)=:\mathcal{C}_i(\boldsymbol{\eta}_{\mathcal{N}(\mathbf{r})}(k)),\,i=1,\ldots,\tilde{b} \quad (2.4)$$

where we define $\mathcal{C}_i$ as the *change in the occupation number* due to interaction. It is given by

2.3. Discrete kinetic theory

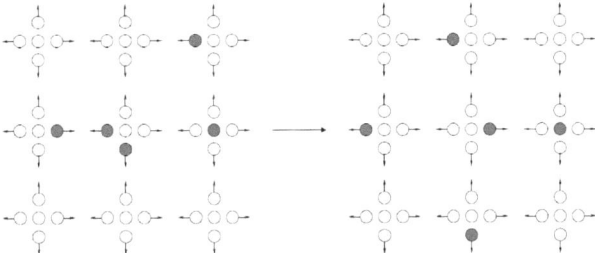

Figure 2.4: Propagation in a two-dimensional square lattice with speed $m = 1$; lattice configurations before and after the propagation step; filled dots denote the presence of a particle in the respective channel.

$$\mathcal{C}_i(\boldsymbol{\eta}_{\mathcal{N}(\mathbf{r})}(k)) = \begin{cases} 1, \text{ creation of a particle in channel } (\mathbf{r}, \mathbf{c}_i) \\ 0, \text{ no change in channel } (\mathbf{r}, \mathbf{c}_i) \\ -1, \text{ annihilation of a particle in channel } (\mathbf{r}, \mathbf{c}_i). \end{cases} \quad (2.5)$$

In a multi-component system with $\sigma = 1, \ldots, \varsigma$ components, eq. (2.4) becomes

$$\eta_{\sigma,i}^{\text{CP}}(\mathbf{r}, k) - \eta_{\sigma,i}(\mathbf{r}, k) = \eta_{\sigma,i}(\mathbf{r} + m_\sigma \mathbf{c}_i, k + \tau) - \eta_{\sigma,i}(\mathbf{r}, k) =: \mathcal{C}_{\sigma,i}(\boldsymbol{\eta}_{\mathcal{N}(\mathbf{r})}(k)), \quad (2.6)$$

for $i = 1, \ldots, \tilde{b}$, with speeds $m_\sigma \in \mathbb{N}_0$ for each component $\sigma = 1, \ldots, \varsigma$. Here, the change in the occupation numbers due to interaction is given by

$$\mathcal{C}_{\sigma,i}(\boldsymbol{\eta}_{\mathcal{N}(\mathbf{r})}(k)) = \begin{cases} 1, \text{ creation of a particle in channel } (\mathbf{r}, \mathbf{c}_i)_\sigma \\ 0, \text{ no change in channel } (\mathbf{r}, \mathbf{c}_i)_\sigma \\ -1, \text{ annihilation of a particle in channel } (\mathbf{r}, \mathbf{c}_i)_\sigma, \end{cases} \quad (2.7)$$

where $(\mathbf{r}, \mathbf{c}_i)_\sigma$ specifies the ith channel associated with node $\mathbf{r}$ of the lattice $\mathcal{L}_\sigma$. In the following, we provide the basic mathematical tools that allows for the analysis of LGCA dynamics.

2.3 Discrete kinetic theory

In the previous section, we have introduced the microdynamical equations of a LGCA. We assume that the automaton's rules are probabilistic (deterministic rules can be considered as a limiting case of probabilistic rules, namely with probabilities of one for certain events). Here, we provide a description of the stochastic process that governs the LGCA evolution. Finally, we develop the formalism of the discrete kinetic theory and, in particular, the derivation of the Lattice Boltzmann Equation (LBE) as a mean-field approximation - defined later in the text - of the LGCA.

2.3.1 Stochastic description

Let us consider the global configuration $\boldsymbol{\eta}(k)$ of the lattice $\mathcal{L}$, which can be thought as a point in a discrete phase space Γ of all $|\Gamma| = 2^{\tilde{b}|\mathcal{L}|}$ possible different microstates. To analyze the behavior of the model, we are interested in the time evolution of an ensemble of microstates. It is convenient to introduce the probability distribution of this phase space $\mathbb{P}(\boldsymbol{\eta}(k))$ for a given time k. We define the ensemble average,

$$\langle ... \rangle = \sum_{\boldsymbol{\eta} \in \Gamma} (...) \mathbb{P}(\boldsymbol{\eta}(k)), \tag{2.8}$$

where the phase space probability distribution is given by

$$\mathbb{P}(\boldsymbol{\eta}(k)) = \left\langle \prod_{\mathbf{r}} \delta(\boldsymbol{\eta}(\mathbf{r}, k)) \right\rangle = \left\langle \prod_{\mathbf{r},i} \delta(\eta_i(\mathbf{r}, k)) \right\rangle, \tag{2.9}$$

where the Dirac function $\delta(\cdot)$ is defined as

$$\delta(x_0) = \begin{cases} 1, & \text{if } x = x_0 \\ 0, & \text{else.} \end{cases}$$

In this thesis, the notations of delta function $\delta(x, x_0)$ or as Kronecker delta δ_{xx_0} will be completely equivalent. The ensemble initial conditions at time $k = 0$ should be endowed with probabilities $\mathbb{P}(\boldsymbol{\eta}(0)) > 0$, such that

$$\sum_{\boldsymbol{\eta} \in \Gamma} \mathbb{P}(\boldsymbol{\eta}(0)) = 1.$$

The entire evolution process of $\{\boldsymbol{\eta}(\cdot, k) : k \in \mathbb{N}\}$ is a stationary *Markov chain*[2] in a discrete phase space Γ, that means the process has no memory. Each element of such ensembles evolves in the phase space Γ via the composition of the automaton's operators. Thus, we can write the evolution equation

$$\mathbb{P}(\boldsymbol{\eta}(\cdot, k + \tau)) = \mathbb{P}(\boldsymbol{\eta}^{\text{CP}}(\cdot, k)). \tag{2.10}$$

The transition from a state $\boldsymbol{\eta}$ to a new state $\tilde{\boldsymbol{\eta}}$ is realized by the probabilities $A_{\boldsymbol{\eta} \to \tilde{\boldsymbol{\eta}}}$, where $\sum_{\boldsymbol{\eta}} A_{\boldsymbol{\eta} \to \tilde{\boldsymbol{\eta}}} = 1$ is the appropriate normalization condition. Now, we can write down the so-called *Chapman-Kolmogorov* equation of the Markov process

$$\mathbb{P}(\boldsymbol{\eta}(\cdot, k + \tau)) = \sum_{\boldsymbol{\eta}, \tilde{\boldsymbol{\eta}}} A_{\boldsymbol{\eta} \to \tilde{\boldsymbol{\eta}}}(\cdot, k) \mathbb{P}(\boldsymbol{\eta}(\cdot, k)), \tag{2.11}$$

---

[2] A stationary Markov process is characterized by time-invariant transition probabilities.

2.3. Discrete kinetic theory

which governs the evolution of probability measures defined in the phase space. Equation (2.11) can be viewed as a *Liouville* equation for the LGCA. Indeed in statistical mechanics the Liouville equation describes the deterministic evolution of a statistical ensemble of initial configurations; in the LGCA models considered here, randomness is also present in the dynamics.

As in statistical mechanics, also in LGCA one faces the complexity of systems with many degrees of freedom. Use of the Chapman-Kolmogorov equation allows one, at least in principle, to recursively determine the evolution of any initial probability measure defined on Γ. In practice this is a formidable task for any system with more than a few nodes. However, full knowledge of the information contained in the probability measure is generally not necessary. The knowledge of the one body reduced distribution function is often sufficient; this observation calls for the elaboration of a theory establishing the evolution of the reduced distribution, bypassing the evaluation of the full probability measure. The next section provides a solution to this problem.

2.3.2 Lattice Boltzmann equations

As stated above, a LGCA can be considered as a system with many degrees of freedom. Here, we present a methodology that allows the reduction of the system's behavior to an effective description with only a few degrees of freedom. The concept of the following methodology has been firstly developed in the pioneering work of Ludwig Boltzmann (1844 – 1906).

The LGCA stochastic process is a stationary Markov chain, i.e. there is no memory effect and the transitions are time-independent. Let us introduce the random variable $\xi_{\boldsymbol{\eta} \to \boldsymbol{\eta}^C}$ which is 1 or 0 depending on whether the collision operator produces the node configuration $\boldsymbol{\eta}^C$ from the initial configuration $\boldsymbol{\eta}$. Now, we can rewrite the microdynamic equations (2.4) as

$$\eta_i(\mathbf{r} + m\mathbf{c}_i, k + \tau) = \sum_{\boldsymbol{\eta}, \boldsymbol{\eta}^C} \eta_i^C(\mathbf{r}, k) \xi_{\boldsymbol{\eta} \to \boldsymbol{\eta}^C} \delta(\boldsymbol{\eta}(\mathbf{r}, k)). \qquad (2.12)$$

The probability that the random variables $\xi_{\boldsymbol{\eta} \to \boldsymbol{\eta}^C}$ take the value 1 is given by

$$\mathbb{P}(\{\xi_{\boldsymbol{\eta} \to \boldsymbol{\eta}^C} = 1\}) = A_{\boldsymbol{\eta} \to \boldsymbol{\eta}^C}(\mathbf{r}, k). \qquad (2.13)$$

Next, we define the *single particle distribution* functions which are the average values of the $\eta_i(\mathbf{r}, k)$, i.e. the average channel occupation number

$$f_i(\mathbf{r}, k) = \langle \eta_i(\mathbf{r}, k) \rangle = \sum_{\boldsymbol{\eta}} \eta_i(\mathbf{r}, k) \mathbb{P}(\boldsymbol{\eta}(\mathbf{r}, k)), \qquad (2.14)$$

where $f_i(\mathbf{r}, k) \in [0, 1] \subset \mathbb{R}$. Now, let's calculate the single particle distribution after the application of C ∘ P:

$$f_i(\mathbf{r} + m\mathbf{c}_i, k + \tau) = \sum_{\boldsymbol{\eta}, \boldsymbol{\eta}^C} \eta_i^C(\mathbf{r}, k) A_{\boldsymbol{\eta} \to \boldsymbol{\eta}^C}(\mathbf{r}, k) \mathbb{P}(\boldsymbol{\eta}(\mathbf{r}, k)). \qquad (2.15)$$

20 Chapter 2. Lattice-gas cellular automata: Basics

The term $\mathbb{P}(\boldsymbol{\eta}(\mathbf{r}, k))$ can be extremely complicated, since the η_i's can be highly correlated. In order to make mathematical analysis feasible, we apply the *mean-field approximation* or *Boltzmann approximation* (Stosszahlansatz), which help us to write down a *completely factorized* node distribution

$$\mathbb{P}(\boldsymbol{\eta}(\mathbf{r}, k)) = \langle \prod_{i=1}^{\tilde{b}} \delta(\eta_i(\mathbf{r}, k)) \rangle = \prod_{i=1}^{\tilde{b}} f_i(\mathbf{r})^{\eta_i(\mathbf{r})} (1 - f_i(\mathbf{r}))^{1-\eta_i(\mathbf{r})}. \qquad (2.16)$$

The mean field approximation discards all the pair or higher on and off node correlations. Now one can derive from the micro-dynamical description (2.4) the mean-field approximation of the LGCA model which is called *non-linear Lattice Boltzmann Equation* (LBE)

$$\boxed{f_i(\mathbf{r} + m\mathbf{c}_i, k + \tau) - f_i(\mathbf{r}, k) = \langle \eta_i^C(\mathbf{r}, k) - \eta_i(\mathbf{r}, k) \rangle =: \tilde{C}_i(\boldsymbol{f}(\mathbf{r}, k))}, \qquad (2.17)$$

where $\boldsymbol{f}(\mathbf{r}, k) = (f_1(\mathbf{r}, k), ..., f_{\tilde{b}}(\mathbf{r}, k))$ and $\tilde{C}_i \in [-1, 1]$ is called *expected collision operator*. Finally, the LBE can be written as the expected dynamics of the stochastic process that governs the LGCA, i.e.

$$f_i(\mathbf{r} + m\mathbf{c}_i, k + \tau) - f_i(\mathbf{r}, k) = \sum_{\boldsymbol{\eta}, \boldsymbol{\eta}^C} (\eta_i^C - \eta_i) A_{\boldsymbol{\eta} \to \boldsymbol{\eta}^C} \prod_{i=1}^{\tilde{b}} f_i(\mathbf{r})^{\eta_i(\mathbf{r})} (1 - f_i(\mathbf{r}))^{1-\eta_i(\mathbf{r})}. \qquad (2.18)$$

In a multi-component system with $\sigma = 1, \ldots, \varsigma$ components, the LBE takes the form

$$f_{\sigma,i}(\mathbf{r} + m_\sigma \mathbf{c}_i, k + \tau) - f_{\sigma,i}(\mathbf{r}, k) = \langle C_{\sigma,i}(\boldsymbol{\eta}(\mathbf{r}, k)) \rangle =: \tilde{C}_{\sigma,i}(\boldsymbol{f}(\mathbf{r}, k)), \qquad (2.19)$$

where $\boldsymbol{f}(\mathbf{r}, k) = (f_{\sigma,1}(\mathbf{r}, k), ..., f_{\sigma,\tilde{b}}(\mathbf{r}, k))_{\sigma=1,\ldots,\varsigma}$ and $\tilde{C}_{\sigma,i} \in [-1, 1]$ the expected collision operator for each species.

2.4 Summary

In this chapter, we presented the lattice-gas cellular automaton's nomenclature and formalism. We introduced the main elements of the discrete kinetic theory that we are going to use in the following chapters. In this context, we derive the so-called lattice Boltzmann equation as the mean-field approximation of LGCA.

Please note that kinetic theory is a physically motivated mathematical theory which describes the spatiotemporal evolution of interacting particles. Therefore, in this chapter, we have used the term particles as a generic term for interacting individuals. In this thesis, our goal is to model and to analyze biological systems that consist of cells. Tumors can be seen as a multi-cellular biological system. In the following chapters, we draw the analogies between the physical multi-particle and multi-cellular systems and we justify the application of discrete kinetic theory.

CHAPTER 3
Random cell motion

Contents

3.1	Introduction	**21**
	3.1.1 Brownian motion	22
	3.1.2 Modeling of Brownian motion	23
3.2	A LGCA model for randomly moving cells	**25**
	3.2.1 Microscopic dynamics	26
	3.2.2 Mesoscopic dynamics	26
3.3	Characterization of random motion	**28**
	3.3.1 Individual cell motion	28
	3.3.2 Collective transport	31
3.4	Summary	**40**

3.1 Introduction

Active migration of blood and tissue cells is essential for a number of physiological processes such as inflammation, wound healing, embryogenesis and tumor cell metastasis [Bray 1992]. Both in natural tissues and artificial environments, such as *in vitro* tissue cultures, cells can exhibit migratory behavior. Cellular motion is a complex phenomenon that involves various intra- and inter-cellular processes. Responding in their surrounding, cells continuously remodel their cytoskeleton and activate a number of molecular motors. The resulting forces alter the cell shape to "pull" the cell body forward [Mogilner 1996]. Thus, cell displacements are a high-level manifestation of a vast array of molecular processes, much like automobile displacements results from a underlying machinery.

Various mathematical models incorporating the above principles of cell motion have been proposed. The most ambitious of them attempt to represent the major physical and chemical processes involved in the motion of an entire cell [Dembo 1989, Dickinson 1993]. Others concentrate on a specific process such as extension of a protrusion [Mogilner 1996] or receptor dynamics [Lauffenburger 1993]. The primary purpose of these biophysical models is to demonstrate that the proposed mechanisms can in fact produce the forces and behaviors observed experimentally.

An alternative approach to cell motion modeling, which is presented in this thesis, is of phenomenological nature. In general, phenomenological models can be used for quantifying the experimentally observed cell behavior without requiring the knowledge of any underlying intracellular mechanisms of cell motion. A phenomenological model of random motion was firstly introduced by the seminal work of R. Brown. In the next sections we introduce history and the different modeling approaches of random Brownian motion. An extension of the Brownian model is the so-called correlated random walks of [Alt 1980, Dunn 1987] and [Shenderov 1997], which have been proposed to describe observations of isolated cells locomoting on a substrate. Diffusive approximations have been widely used to describe the behavior of moving cell populations following [Keller 1971]. The theoretical relationships between single cell models and population models have been studied in numerous works [Alt 1980, Dickinson 1993, Ford 1991].

Here, we present a LGCA model for random cell motion. This model mimics the phenomenology of a randomly moving single cell. In addition, our model defines a stochastic process that provides a mesoscopic description of the random motion of a "small" ensemble of cells. The model allows for the prediction of the dynamics of the whole population of randomly moving cells. The biophysical details of cell motion are beyond the model's resolution. The classical physical prototype model related to our LGCA model is the so-called Brownian motion, which is described below. Finally, in order to characterize the random cell motion, we calculate the motility rate of a single cell or an ensemble of cells. In particular, in the latter case we introduce three *scaling methods* that allow for the transition from microscopic to macroscopic dynamics. However, we do not provide any exhaustive comparison of these methods since it is out of the scope of this thesis. These scaling methods will be also used in the following chapters of this thesis.

3.1.1 Brownian motion

In 1827 Robert Brown, a well-known botanist, was studying sexual relations of plants, and in particular was interested in the particles contained in grains of pollen. He began with a plant (*Clarckia pulchella*) in which he found the pollen grains were filled with oblong granules about 5 microns long. He noticed that these granules were in constant motion, and satisfied himself that this motion was not caused by currents in the fluid or evaporation. Smaller spherical grains had even more vigorous motion. He thought at first that he was looking at the plant equivalent of sperm they were jiggling around because they were alive. To check this, he repeated the same experiment with dead plants and he found that there was just as much jiggling. Perhaps all organic matter, everything that ever was alive, still contained some mysterious life force at this microscopic level? Sure enough, he found the movement in tiny fragments of fossilized wood! But then he went on to find it in matter that never was alive tiny particles of window glass, and even dust from a stone that had been part of the Sphinx. The movement evidently had nothing to do with the substance ever being alive or dead, much to Brown's surprise. So what was

3.1. Introduction

causing it? Perhaps it was evaporation currents, or the incident light energy, or just tiny unnoticed vibrations. But none of these explanations was very satisfactory.

Half a century later, a new possible explanation emerged. The kinetic theory of heat developed by Maxwell, Boltzmann and others was gaining credence. If all the molecules in the fluid were indeed in vigorous motion, maybe these tiny granules were being moved around by this constant battering from all sides as the fluid molecules bounced off. But there was a problem with this explanation: didn't it violate the second law of thermodynamics? It had been well established that energy always degrades, as friction slows movement kinetic energy goes to heat energy. This seemed to be the other way round the molecular battering was certainly disorganized heat energy, but when the granule moved it had evidently gained kinetic energy. Since many scientists regarded the second law as an absolute truth, they were very skeptical of this explanation.

In 1888, French experimentalist Leon Gouy investigated the particle movement in detail, finding it to be more lively in low viscosity liquids. He established that it was unaffected by intense illumination or by strong electromagnetic fields. Despite the second law, Guoy believed – correctly – the random motion was indeed generated by thermal molecular collisions [1].

3.1.2 Modeling of Brownian motion

The goal of this subsection is to characterize mathematically Brownian motion. In particular, we use the mathematical formulation of the problem proposed by Einstein in his fundamental paper in 1905 [Einstein 1905]. In this paper, Einstein undertakes a probabilistic derivation of the diffusion equation based on a particularly simple model for the fluctuations of the Brownian particle, called a *random walk*. The essential idea is to express the probability $p(x, t + \tau)$ (in one space dimension) of finding a *single particle* at position x and time $t + \tau$ as a function of the probability at time t,

$$p(x, t+\tau) = \int_{\xi \in \mathbb{R}} d\xi \, p(x+\xi, t) \, \phi(\xi), \qquad (3.1)$$

where ξ is the length of a particle jump. Due to the use of the auxiliary "jump probabilities" $\phi(\xi)$ (probability to jump to the left or to the right) in the master equation (3.1) (called nowadays Chapman-Kolmogorov equation), Einstein's derivation appears much more elegant and transparent than the one proposed by Smoluchowski, that includes a more mechanistic description of Brownian motion [Smoluchowski 1906]. Einstein's approach circumvents the construction of an explicit *microscopic dynamical theory* of the process by some (implicit) technical assumptions on $\phi(\xi)$. An important assumption is the symmetry property $\phi(+\xi) = \phi(-\xi)$ which implies that the probability of a jump of length ξ to the left or to the right are equal. Moreover, since $\phi(\xi)$ functions represent probabilities then the following normalization relation should be satisfied:

---
[1]The information about the history of Brownian has been retrieved from [Magie 1963]

$$\int_{\xi \in \mathbb{R}} \phi(\xi) \, d\xi = 1. \tag{3.2}$$

From eq. (3.1) we can obtain an evolution equation for the particle concentration $\rho(x,t) = \int_0^1 n\, p(x,t)\, dn$, which is the normalized particle density at a site x at time t (it can be seen also as the probability of finding a particle at a site x at time t):

$$\int_0^1 n\, p(x, t+\tau)\, dn = \int_0^1 \int_{\xi \in \mathbb{R}} n\, p(x+\xi, t)\, \phi(\xi)\, d\xi\, dn$$

$$\Rightarrow \rho(x, t+\tau) = \int_{\xi \in \mathbb{R}} d\xi \, \rho(x+\xi, t)\, \phi(\xi), \tag{3.3}$$

where the above equation is a Coupled Map Lattice (CML). Taylor expanding $\rho(x,t)$ with respect to τ and ξ in the eq. (3.3) yields

$$\rho(x,t) + \tau \frac{\partial \rho(x,t)}{\partial t} = \rho(x,t) \int_{\xi \in \mathbb{R}} \phi(\xi)\, d\xi +$$
$$\frac{\partial \rho(x,t)}{\partial x} \int_{\xi \in \mathbb{R}} \xi\, \phi(\xi)\, d\xi + \frac{\partial^2 \rho(x,t)}{\partial x^2} \int_{\xi \in \mathbb{R}} \frac{\xi^2}{2} \phi(\xi)\, d\xi. \tag{3.4}$$

Using the symmetry property $\phi(+\xi) = \phi(-\xi)$ and eq. (3.2), eq. (3.4) becomes

$$\tau \frac{\partial \rho(x,t)}{\partial t} = \frac{\partial^2 \rho(x,t)}{\partial x^2} \int_{\xi \in \mathbb{R}} \frac{\xi^2}{2} \phi(\xi)\, d\xi. \tag{3.5}$$

At this point Einstein has achieved a description for the diffusion coefficient based on microscopic statistical observables, such as the mean square displacement of a *single particle* $\langle X^2 \rangle$. In particular, Einstein's formula for diffusion in one dimension yields

$$D^* = \frac{1}{2\tau} \int_{\xi \in \mathbb{R}} \xi^2\, \phi(\xi)\, d\xi = \frac{\langle X^2 \rangle}{2\tau}, \tag{3.6}$$

Note that the above expression is valid only if the observation time interval τ is much larger than the microscopic collision time. Replacing eq. (3.6) in eq. (3.5)

$$\frac{\partial \rho(x,t)}{\partial t} = D^* \nabla^2 \rho(x,t), \tag{3.7}$$

which coincides with the famous diffusion equation.

Now, we can calculate the mean square displacement of an *ensemble* of particles directly from the diffusion equation (3.7) which describes the ensemble's spatiotemporal evolution. If we multiply eq. (3.7) by x^2, integrate over the whole domain $\mathbb{R}$ and assume that the density and its spatial derivatives tend to zero when $|x| \to \infty$, we obtain

$$\frac{d\langle x^2 \rangle}{dt} = 2D, \qquad (3.8)$$

where $\langle x^2 \rangle = \int_{\mathbb{R}} x^2 \rho(x,t) dx$ is the mean square displacement of the ensemble. If we assume that $\rho(x,0) = \delta(x)$ and $\partial_x \rho(x,0) = 0$, then the diffusion coefficient reads

$$D = \frac{\langle x^2 \rangle}{2t}. \qquad (3.9)$$

We observe that for $t = \tau$ the definitions of the diffusion coefficients D and D^* are identical. This issue will be discussed in the next sections, since there are cases in which the two diffusion coefficients differ.

The work of Einstein has shown that kinetic coefficients (here diffusion coefficient) in macroscopic equations (here diffusion equation) can be expressed entirely in terms of correlation functions of fluctuations of microscopic variables. This is how a discrete, chaotic processes on the microscale (random particle collisions) generate smooth behavior on the macroscopic scale.

3.2 A LGCA model for randomly moving cells

We construct a model system with stochastic microdynamics which, at the mesoscopic level, can approximately describe the phenomenology of randomly moving cells. In our model, cell dynamics are identified by the LGCA rules. Automaton dynamics arise from the repetition of two rules (*operators*): Propagation (P) and reorientation (O).

The propagation operator (P) was defined in detail in the previous chapter. Briefly, each cell in the automaton moves to the same channel of a neighboring node according to its velocity. For instance, if there is a cell in the channel "up" of node $\mathbf{r}$, the propagation step will move it to the channel "up" of the node $\mathbf{r} + \mathbf{c}_2$, where $\mathbf{c}_2$ is the unit vector (0,1) (for details see the previous chapter).

The reorientation operator is responsible for the redistribution of cells within the velocity channels of a node, providing a new node velocity distribution (see fig. 3.1). In this chapter, we assume that individual cells perform random walks. The corresponding transition probabilities are

$$\mathbb{P}(\boldsymbol{\eta} \to \boldsymbol{\eta}^O)(\mathbf{r},k) = \frac{1}{Z} \delta\big(n(\mathbf{r},k), n^O(\mathbf{r},k)\big), \qquad (3.10)$$

where the normalization factor $Z = \sum_{\boldsymbol{\eta}^O(r,k)} \delta\big(n(\mathbf{r},k), n^O(\mathbf{r},k)\big)$ corresponds to the equivalence class defined by the value of the pre-interaction node density $n(\mathbf{r},k)$.

Obviously, this case implies a uniformly random redistribution of the cells among the node's channels. The Kronecker δ assumes the mass conservation of this operator. This choice of the reorientation operator is one out of the possible ways to describe random motion by means of LGCA proposed in [Chopard 1998, Deutsch 2005]. However, this rule greatly simplifies the subsequent analytical derivation of the equations describing the macro- and mesoscopic evolution of the automaton.

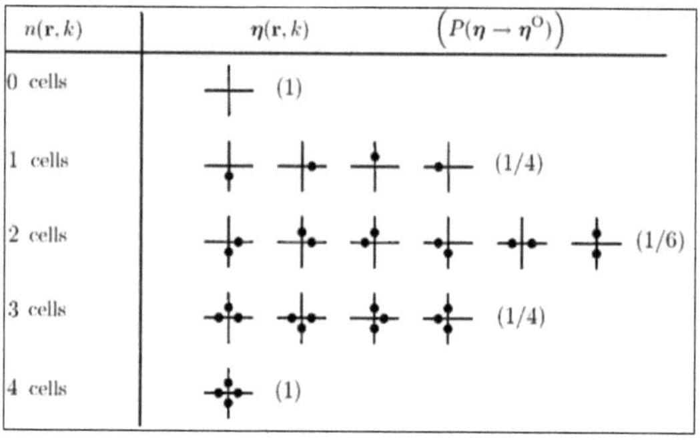

26   Chapter 3. Random cell motion

$n(\mathbf{r},k)$	$\eta(\mathbf{r},k)$	$\left(P(\eta \to \eta^O)\right)$
0 cells	─┼─ (1)	
1 cells	─┼ ─┼─ ─┼─ ┼─ (1/4)	
2 cells	─┼ ─┼─ ─┼─ ┼─ ─┼─ ┼─ (1/6)	
3 cells	─┼─ ─┼─ ─┼─ ─┼─ (1/4)	
4 cells	─┼─ (1)	

Figure 3.1: Reorientation rule of random motion. The first column corresponds to the number of cells on a node $n(\mathbf{r}, k)$ at a time $k m$, with capacity $\tilde{b} = 4$. The right column indicates all the possible cell configurations on node and the transition probability of obtaining a certain configuration (3.10).

3.2.1 Microscopic dynamics

The temporal dynamics of our LGCA are provided by the composition of propagation (P) and the reorientation (O) operators. The corresponding microdynamical equations are given by:

$$\eta_i(\mathbf{r} + m\mathbf{c}_i, k + \tau) = \sum_{\eta, \eta^O} \eta_i^O(\mathbf{r}, k) \xi_{\eta \to \eta^O} \delta(\eta(\mathbf{r}, k)). \qquad (3.11)$$

Eq. (3.11) refers to the *redistribution* of cells on the velocity channels and the *propagation* to the neighboring nodes. The random variable $\xi_{\eta \to \eta^O}$ is 1 or 0 depending on whether the reorientation operator produces the node configuration η^O from the initial configuration η.

3.2.2 Mesoscopic dynamics

We introduce the *single particle distribution* functions which are the average values of the $\eta_i(\mathbf{r}, k)$, i.e. the average channel occupation number

$$f_i(\mathbf{r}, k) = \sum_{\eta} \eta_i(\mathbf{r}, k) \mathbb{P}(\eta(\mathbf{r}, k)). \qquad (3.12)$$

The probability that the random variables $\xi_{\eta \to \eta^O}$ take the value 1 is given by

$$\mathbb{P}(\{\xi_{\eta \to \eta^O} = 1\}) = A_{\eta \to \eta^O}(\mathbf{r}, k), \qquad (3.13)$$

which are the transition probabilities defined in (3.10).

3.2. A LGCA model for randomly moving cells 27

Now, let us calculate the single particle distribution after the application of P∘O:

$$f_i(\mathbf{r} + m\mathbf{c}_i, k + \tau) = \sum_{\boldsymbol{\eta},\boldsymbol{\eta}^\circ} \eta_i^O(\mathbf{r}, k) A_{\boldsymbol{\eta} \to \boldsymbol{\eta}^\circ}(\mathbf{r}, k) \mathbb{P}\big(\boldsymbol{\eta}(\mathbf{r}, k)\big). \qquad (3.14)$$

In the appendix of this chapter, we calculate in detail the exact expression of the r.h.s of the above equation. Using the appendix results

$$\sum_{\boldsymbol{\eta}^\circ} f_i^O(\mathbf{r}, k) A_{\boldsymbol{\eta} \to \boldsymbol{\eta}^\circ}(\mathbf{r}, k) = \frac{1}{\tilde{b}} \sum_{i=1}^{\tilde{b}} f_i(\mathbf{r}, k), \qquad (3.15)$$

we can obtain the Lattice Boltzmann Equation (LBE) which reads:

$$f_i(\mathbf{r} + m\mathbf{c}_i, k + \tau) - f_i(\mathbf{r}, k) = \sum_{j=0}^{\tilde{b}} \Omega_{ij} f_j(\mathbf{r}, k), \qquad (3.16)$$

where the matrix $\Omega = (1/\tilde{b} - \delta_{ij})_{i,j=1,\ldots,\tilde{b}}$ is the transition matrix of the underlying shuffling process. The transition matrix defines the probabilities of a channel changing its state. Note that, due to the simplicity of the process, the above Boltzmann equation is *exact*, i.e. it can be obtained without using additional assumptions (e.g. the mean-field assumption).

Now we define the the *mean node density* as

$$\rho(\mathbf{r}, k) = \langle n(\mathbf{r}, k) \rangle = \sum_{i=1}^{\tilde{b}} f_i(\mathbf{r}, k). \qquad (3.17)$$

Summing the LBE (3.16) over the number of channels and changing the variables as $\mathbf{r} = \mathbf{r} - m\mathbf{c}_i$, we obtain:

$$\sum_{i}^{\tilde{b}} f_i(\mathbf{r}, k + \tau) = \frac{1}{\tilde{b}} \sum_{i,j=1}^{\tilde{b}} f_j(\mathbf{r} - m\mathbf{c}_i, k) = \frac{1}{\tilde{b}} \sum_{i=1}^{\tilde{b}} \rho(\mathbf{r} - m\mathbf{c}_i, k), \qquad (3.18)$$

Now the LBE (3.16) can be rewritten in terms of average node densities:

$$\rho(\mathbf{r}, k + \tau) = \frac{1}{\tilde{b}} \sum_{i=1}^{\tilde{b}} \rho(\mathbf{r} - m\mathbf{c}_i, k). \qquad (3.19)$$

Ths equation is equivalent to a mesoscopic Coupled Map Lattice (CML) model for diffusion. We will see that the above formulation strongly facilitates the following analytical manipulations.

3.3 Characterization of random motion

In this chapter, the main assumption is that cell random motion can be identified by random walk. In order to characterize cell motility, one should devise the appropriate statistical observables. An intuitive guess would be the mean cell velocity. As shown below, the mean cell velocity is always zero for randomly moving cells, which means that we need another observable to characterize the random cell motion. A good candidate can be considered the mean square displacement of a cell as a function of time (see sec. 3.1.2). In the following, we demonstrate how for our LGCA the mean square displacement is closely related to the definition of the diffusion coefficient, which usually characterizes cell motility. Since cell motion is dominated by randomness, the mean square displacement is a random variable and may be only specified in terms of probability distributions. In the following analysis, we assume that the probability distributions, related to lattice entities (nodes, channels etc.), are stationary. Therefore, the following analytical calculations are conducted under the assumption that the node probability distribution relaxes to an equilibrium, i.e. time-invariant distribution. Finally, we study the motility properties in the case of (i) individual cell motion and of (ii) cell ensemble dispersion. The reason for distinguishing these two cases is due to the fact the motility of a single cell and of a cell ensemble can differ from each other. A striking example is the case of an LGCA with two sites and full with cells. The cells can exchange constantly nodes, i.e. their motility rate is different from zero, but the population appears to be immobile. This phenomenon is discussed in detail in [Philibert 1990].

3.3.1 Individual cell motion

Our LGCA model provides a stochastic description of unbiased random cell motion, as introduced in sec. 3.1.2. To analyze the properties of cell motility, we observe the dynamics of a single cell. The statistical observable of interest is the cell displacement $\Delta x(k)$,

$$\Delta x(k) = x(k) - x(0) = \sum_{\delta t=0}^{k-1} v(\delta t), \qquad (3.20)$$

where $v(k)$ is the velocity at a discrete time k of a single cell which maybe expressed in terms of the occupation numbers $\eta_i(\mathbf{r}, k)$, i.e.

$$v(k) = \sum_{\mathbf{r},i} m\mathbf{c}_i \eta_i(\mathbf{r}, k), \qquad (3.21)$$

where $\sum_{\mathbf{r},i} \eta_i(\mathbf{r}, k) = 1$.

The trivial equilibrium probability distribution[2] of finding an individual cell in channel i is a Bernoulli distribution, where $f_i(\mathbf{r}, k) = \langle \eta_i(\mathbf{r}, k) \rangle_{eq} = \mathbb{P}\{\eta_i(\mathbf{r}, k) = 1\}$

---

[2] Due to the simplicity of the orientation operator, the equilibrium distribution of the occupation numbers is attained after a single simulation step.

3.3. Characterization of random motion

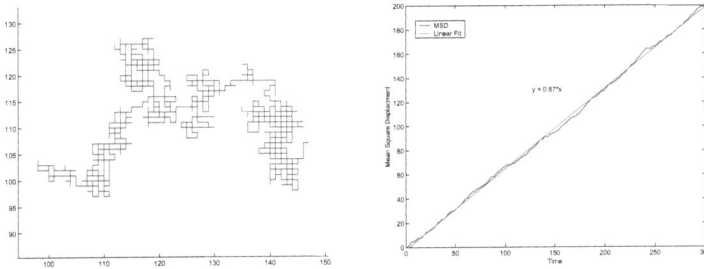

Figure 3.2: The left figure presents the trajectory of a randomly moving cell. The right figure shows the time evolution of the corresponding mean square displacement. The lattice consist of nodes with $\tilde{b} = 6$. The expected diffusion coefficient of an individual cell is $D^* = \Delta \langle X_k^2 \rangle / b = 0.67/4 \simeq 1/6$, where $b = 4$ the dimension of the system.

is the *single particle distribution* or mean occupation number of a channel at a node $\mathbf{r}$ and time k. It is easy to deduce that for a single single cell f_i is equal to the constant value $1/\tilde{b}$.

The mean cell displacement X_k, at time k, in the equilibrium ensemble is given by:

$$\langle X_k \rangle_{eq} = \langle \sum_{\delta t=0}^{k-1} \Delta x(\delta t) \rangle_{eq} = \sum_{\delta t=0}^{k-1} \sum_{\mathbf{r},i} m \mathbf{c}_i \langle \eta_i(\mathbf{r},\delta t) \rangle_{eq} = \sum_{\delta t=0}^{k-1} \sum_{\mathbf{r},i} m \mathbf{c}_i f_i(\mathbf{r},\delta t) = \mathbf{0}, \quad (3.22)$$

due to the symmetry property of the square lattice tensor $\sum_{i=1}^{\tilde{b}} \mathbf{c}_i = \mathbf{0}$. Thus, the absolute mean displacement of a randomly moving cell is always zero due to the equiprobable choice of directions. Then we choose to observe the mean square displacement $\langle X_k^2 \rangle$ of a cell, i.e.

$$\langle X_k^2 \rangle_{eq} = \sum_{\delta t_1=0}^{k-1} \sum_{\delta t_2=0}^{k-1} \langle \Delta x(\delta t_1) \Delta x(\delta t_2) \rangle_{eq} \quad (3.23)$$

The equilibrium covariance matrix is $C_{nm} = \langle \Delta x(n) \Delta x(m) \rangle_{eq}$, where $n, m \in \mathbb{N}$ is a symmetric matrix with non-zero elements along the diagonal. Thus, eq. (3.23) can be rewritten as

$$\langle X_k^2 \rangle_{eq} = \sum_{\delta t_1=0}^{k-1} \sum_{\delta t_2=0}^{k-1} C(|\delta t_1 - \delta t_2|) \quad (3.24)$$

The sum (3.24) has k^2 terms; we group terms with $|\delta t_1 - \delta t_2| = \delta t \neq 0$ and observe that the number of such terms is $2(k - \tau)$. Then we notice that the number of terms $\delta t_1 = \delta t_2$ is k and that $C(0) = \langle \Delta x^2 \rangle_{eq}$ and obtain:

$$\langle X_k^2 \rangle_{eq} = kC(0) + 2 \sum_{\delta t=0}^{k-1} (k - \delta t) C(\delta t), \quad (3.25)$$

where $\delta t = |\delta t_1 - \delta t_2|$. Combining equations (3.20, 3.21, 3.25) we obtain the famous *Green-Kubo* formula [Kubo 1980]:

$$\langle X_k^2 \rangle_{eq} = k \langle v^2(0) \rangle_{eq} + 2 \sum_{\delta t=0}^{k-1} (k - \delta t) \langle v(0) v(\delta t) \rangle_{eq}. \quad (3.26)$$

The quantity $\langle v(0) v(\tau) \rangle$ is called *velocity autocorrelation* function. In our model, the autocorrelation function is zero, since the correlation function $C_{nm} = 0$ (note that for the square lattice $\sum_{i,j} \mathbf{c}_i \mathbf{c}_j = 0$ for $i \neq j$).

From the famous Einstein formula (3.6), we can calculate the diffusion coefficient as the rate of change of the mean square displacement of a single cell, i.e.

$$D^* = \frac{\langle X_k^2 \rangle_{eq}}{2dk\tau}, \quad (3.27)$$

where d is the dimension of the space (in our case $d = 2$). Please note, that the Einstein formula indicates that for a random walk process the ratio between mean square displacement and time is constant. Thus, specifically in our case, using the Green-Kubo formula, we obtain:

$$2d\tau D^* = \langle v^2(0) \rangle_{eq} = \sum_{\mathbf{r},i} m^2 \mathbf{c}_i^2 \langle \eta_i^2(r,0) \rangle_{eq} = m^2 \sum_{\mathbf{r},i} \mathbf{c}_i^2 \langle \eta_i(r,0) \rangle_{eq} = b \frac{m^2}{\tilde{b}}. \quad (3.28)$$

Here we have used the fact that $\eta_i(\mathbf{r},0) \sim Bernoulli(f_i = 1/\tilde{b})$, therefore $\langle \eta_i^2(\mathbf{r},0) \rangle = \langle \eta_i(\mathbf{r},0) \rangle = 1/\tilde{b}$. Since the coordination number is $b = 2d$, we obtain the following theoretical value for the diffusion coefficient of a single cell:

$$D^* = \frac{m^2}{\tilde{b}\tau}. \quad (3.29)$$

The result (3.29) is confirmed by the simulations of our LGCA. In particular, we simulate our system in a square lattice with $\beta = 2$ rest channels, i.e. $\tilde{b} = 6$ for 300 time steps and we observe the time evolution of the mean square displacement of a single cell (Fig. 3.2). The diffusion coefficient will be given by the formula $D^* = \Delta \langle X_k^2 \rangle / b$, where $\Delta \langle X_k^2 \rangle$ denotes the slope of the curve and $b = 4$. The calculated value is $D^* \simeq 0.167$ which coincides with the theoretical value given by eq. (3.29), which $m = 1$ as in the simulations.

3.3. Characterization of random motion

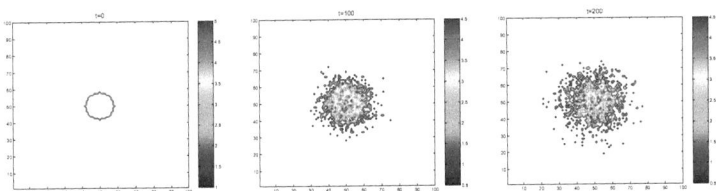

Figure 3.3: Temporal evolution of a diffusive cell cluster at times $t = 0, 100, 200$. The initial configuration represents a fully occupied circle of nodes. Colors indicate the node density (see colorbar).

3.3.2 Collective transport

In the previous section, we have discussed the motion of a single cell and we have calculated the single cell motility rate. In this subsection, our focus is on the calculation of the collective motility rate of a cell population. In fig. 3.3, we observe the typical temporal evolution of an ensemble randomly moving cells.

To characterize the collective motion of randomly moving cells, we need to derive the macroscopic dynamics of the system. We present three different methods that allow for the transition from a microscopic to a macroscopic description: the Chapman-Enskog method, the limit of a CML (3.19) and the Fourier space method. In all three cases the resulting macroscopic description coincides with the diffusion equation. Moreover, we can derive another limiting equation of the LGCA, the so-called telegraph equation, which describes the short-term dynamics of a cell population motion. Finally, we compare the analytically calculated motility rates with the simulation results.

3.3.2.1 Chapman-Enskog method

The Chapman-Enskog method for LGCA is described in detail in [Chopard 1998]. In the following, we equip the parabolic (or diffusive) spatio-temporal scaling which provides the long-term macroscopic dynamics for a diffusion-based process, i.e.

$$\mathbf{x} = \varepsilon \mathbf{r} \text{ and } t = \varepsilon^2 k, \quad (3.30)$$

where $(\mathbf{x}, t)$ are the continuous variables as $\varepsilon \to 0$. The small parameter ε is defined as the ratio of the microscopic mean free path, i.e. the microscopic jump m of a cell within a microscopic time step τ, and the macroscopic length scale, i.e. the length of the domain L:

$$\varepsilon = \frac{m}{L} \ll 1. \quad (3.31)$$

The above definition of ε coincides with the definition of the *Knudsen number* in classical kinetic theory. The physical motivation for the diffusive scaling stems from

the Einstein formula for diffusion (3.6), where the ratio between the squared spatial increment (mean square displacement) and the time step should be constant.
Replacing the first part of equation (3.16) by its Taylor expansion leads to:

$$f_i(\mathbf{r} + m\mathbf{c}_i, k + \tau) - f_i(\mathbf{r}, k) = \left(\varepsilon^2 \tau \partial_t + \varepsilon^4 \frac{\tau^2}{2} \partial_{tt} + \varepsilon m(\mathbf{c}_i \cdot \nabla)\right) \quad (3.32)$$
$$+ \varepsilon^2 \frac{m^2}{2}(\mathbf{c}_i \cdot \nabla)^2 + \varepsilon^3 \tau m \partial_t(\mathbf{c}_i \cdot \nabla)\right) f_i(\mathbf{r}, k).$$

Furthermore, we assume an asymptotic solution of the single particle distribution in the form of

$$f_i(\mathbf{r}, k) = f_i^{(0)} + \varepsilon^1 f_i^{(1)} + \varepsilon^2 f_i^{(2)} + \mathcal{O}(\varepsilon^3).$$

The next step is to insert eq. (3.32) into (3.16) and to collect the terms of equal ε order:

$$\mathcal{O}(\varepsilon^0) \; : \; \sum_j \Omega_{ij} f_j^{(0)} = 0, \quad (3.33)$$

$$\mathcal{O}(\varepsilon^1) \; : \; m(\mathbf{c}_i \cdot \nabla) f_j^{(0)} = \sum_j \Omega_{ij} f_j^{(1)}, \quad (3.34)$$

$$\mathcal{O}(\varepsilon^2) \; : \; \tau \partial_t f_i^{(0)} + m(\mathbf{c}_i \cdot \nabla) f_i^{(1)} + \frac{m^2}{2}(\mathbf{c}_i \cdot \nabla)^2 f_i^{(0)} = \sum_j \Omega_{ij} f_j^{(2)}. \quad (3.35)$$

The macroscopic quantity of interest is the cell density $\rho = \sum_i f_i^{(0)}$, with the assumption $\sum_i f_i^{(l)} = 0$, if $l \geq 1$ [Hillen 2006] (this means that the only the zeroth order contributes to the density). The above relations can also be seen as the solvability conditions for the solution of the following system of equations (discrete Fredholm alternative [Elton 1995]). The solutions of the above equations (3.33), (3.34), (3.35) are based on the properties of the transition matrix Ω [Chopard 1998] and the results are:

$$f_j^{(0)} = \frac{\rho}{\tilde{b}}, \quad (3.36)$$

$$f_j^{(1)} = -\frac{m}{\tilde{b}}(\mathbf{c}_i \cdot \nabla)\rho. \quad (3.37)$$

To obtain a macroscopic equation that describes the spatiotemporal evolution of the cell density, we sum equation (3.35) over i; it follows:

$$\tau \partial_t \sum_i f_i^{(0)} + m \sum_i (\mathbf{c}_i \cdot \nabla) f_i^{(1)} + \frac{m^2}{2} \sum_i (\mathbf{c}_i \cdot \nabla)^2 f_i^{(0)} = \sum_{ij} \Omega_{ij} f_j^{(2)}. \quad (3.38)$$

Performing the above summation (using the property of the transition matrix $\sum_{ij} \Omega_{ij} = 0$ and the property of the lattice tensor $\sum_i c_{i\alpha} c_{i\beta} = d\delta_{\alpha\beta}$, where d is the dimension of the system), yields the following diffusion equation (3.7):

$$\partial_t \rho = \frac{m^2}{\tau \tilde{b}} \nabla^2 \rho, \quad (3.39)$$

where the diffusion coefficient $D = m^2/\tau \tilde{b}$ is finite as $\varepsilon \to 0$.

3.3.2.2 Limit of the CML equation

Now, we rewrite eq. (3.19) as

$$\rho(\mathbf{r}, k+\tau) = \frac{1}{\tilde{b}} \sum_{i=1}^{\tilde{b}} \rho(\mathbf{r} - m\mathbf{c}_i, k)$$

$$\Rightarrow \rho(\mathbf{r}, k+\tau) = (1-\mu)\rho(\mathbf{r},k) + \frac{\mu}{b}\sum_{i=1}^{b} \rho(\mathbf{r} - m\mathbf{c}_i, k), \quad (3.40)$$

where $\mu = \frac{b}{\tilde{b}}$ is the fraction of cells that remain at the current node. The above CML implies that the new node density is given by the average density of the local node neighborhood.

As shown in [White 2005], CML models can be translated into an integro-difference equation such as:

$$\rho(\mathbf{r}, k+\tau) = \int_{\mathcal{L}} \rho(\mathbf{r} - m\mathbf{c}_i, k)\phi(m)dm, \quad (3.41)$$

where $\phi(+m) = \phi(-m)$ is the isotropic (symmetric) redistribution kernel of the process, which represents the rate that a cell makes a jump of length m. As we have seen in the subsect. 3.1.2, eq. (3.41) represents the averaged continuous Chapman-Kolmogorov equation (master equation) of the underlying stochastic process. In our case, the discrete redistribution kernel has the form of:

$$\phi(m) = (1-\mu)\delta(\mathbf{r}) + \frac{\mu}{b}\sum_{i=1}^{b}\delta(\mathbf{r} - m\mathbf{c}_i). \quad (3.42)$$

Since these kernels should be mass conserving, the condition $\int_{\mathcal{L}} \phi(m)dm = 1$ has to be satisfied. Integro-difference equations have been studied extensively in mathematical, biological and specially in ecological literature [Kot 1992, Neubert 1995, Medlock 2003]. Therefore, there exists a vast repertoire of tools that allows for their mathematical analysis.

Here we derive the macroscopic limit of eq. (3.40) using the same parabolic scaling argument as in the Chapman-Enskog method. Expanding the CML eq. (3.40) in a Taylor series for $t \ll \tau$ and $x \ll m$, we obtain the equation:

$$\sum_{n=1}^{\infty}\frac{\tau^n}{n!}\partial_t\rho(x,t) = \sum_{n=1}^{\infty}\frac{m^{2n}}{n!\tilde{b}}\partial_x^{2n}\rho(x,t). \quad (3.43)$$

Up to the first order in τ and up to second order in m, this yields the classical diffusion equation (3.7):

$$\partial_t \rho = D\nabla^2 \rho, \quad (3.44)$$

with diffusion coefficient $D = m^2/\tau\tilde{b}$, which is identical to the macroscopic description (3.52) derived by means of Chapman-Enskog method.

3.3.2.3 Fourier space method

The lattice Boltzmann equation of our LGCA defines a system of equations that describes the spatio-temporal evolution for the vector $\mathbf{f}(\mathbf{r}, k) = (f_1, ..., f_{\tilde{b}})(\mathbf{r}, k)$

$$\mathbf{f}(\mathbf{r} + m\mathbf{c}_i, k + \tau) = \mathbf{\Gamma}\mathbf{f}(\mathbf{r}, k), \tag{3.45}$$

where $\mathbf{\Gamma} = \mathbf{I} + \mathbf{\Omega}$ is the time-domain *Boltzmann propagator* with dimensions $\tilde{b} \times \tilde{b}$ and $\Omega_{ij} = 1/\tilde{b} - \delta_{ij}$, for $i, j = 1, ..., \tilde{b}$. Now, since the system (3.45) is linear, we can introduce the two-dimensional, discrete Fourier transform with wavenumber $\mathbf{q} = (q_1, q_2)$ of the corresponding Fourier mode:

$$f_i(\mathbf{r}, k) = A^{\tau k} e^{i\langle \mathbf{q}, \mathbf{c}_i \rangle m} \hat{f}_i, \tag{3.46}$$

where $\langle \cdot, \cdot \rangle$ is the inner product of two vectors. Then from the system (3.45), we obtain the following algebraic set of equations for the $\hat{f}_i$'s:

$$\mathbf{M}\hat{\mathbf{f}} = 0, \tag{3.47}$$

where the matrix $\mathbf{M}$ is the Fourier transform of the Boltzmann propagator $\mathbf{\Gamma}$ and $\hat{\mathbf{f}} = (\hat{f}_1, ..., \hat{f}_{\tilde{b}})$. The elements of $\mathbf{M}$ are:

$$\mathbf{M}_{ij} = \frac{1}{\tilde{b}} - A^\tau e^{i\langle \mathbf{q}, \mathbf{c}_i \rangle m} \delta_{ij}, \tag{3.48}$$

where $i, j = 1, ..., \tilde{b}$. A non-trivial solution exists only if $det(\mathbf{M}) = 0$. Making explicit use of this condition, we obtain a $\tilde{b}^{th}$ order polynomial equation for the damping coefficient A

$$\frac{2}{\tilde{b}} A^{\tau(\tilde{b}-1)}[\cos(mq_1) + \cos(mq_2) + \frac{\tilde{b}}{2}A + \frac{\tilde{b}}{2} - 2] = 0. \tag{3.49}$$

The solutions of A for the above discrete dispersion relation are:

$$A^\tau_{(1)}(\mathbf{q}) = 2\frac{\cos(mq_1) + \cos(mq_2)}{\tilde{b}} + 1 - \frac{4}{\tilde{b}},$$
$$A^\tau_{(j)}(\mathbf{q}) = 0, \text{ for } j = 2, ..., \tilde{b}.$$

The damping coefficient $A^\tau_{(1)}$ depends on $\mathbf{q}$ and its value is different from zero. For small wavenumbers $|\mathbf{q}| \to 0$, we can write the Taylor expansion of the damping coefficient $A^\tau_{(1)}$:

$$A^\tau_{(1)}(\mathbf{q}) = 1 - \frac{m^2}{\tilde{b}}|\mathbf{q}|^2 + \mathcal{O}(\mathbf{q}^4) \tag{3.50}$$

In the limit of $|\mathbf{q}| \to 0$ the maximum damping factor $A^\tau_{(1)} \to 1$, i.e. this mode is neutrally stable. Moreover, the leading damping coefficient can be expressed as the exponential of an equivalent continuous damping rate $z(\mathbf{q})$, i.e. $A^\tau_{(1)}(\mathbf{q}) = e^{z(\mathbf{q})}$ or $z(\mathbf{q}) = \tau ln(A_{(1)}(\mathbf{q}))$. After this transformation the eq. (3.50) yields

3.3. Characterization of random motion

$$z(\mathbf{q}) = -\frac{m^2}{\tilde{b}\tau}|\mathbf{q}|^2 + \mathcal{O}(\mathbf{q}^4). \tag{3.51}$$

The above equation coincides with the dispersal equation (or characteristic equation) of the diffusion equation (3.7):

$$\partial_t \rho = D\nabla^2 \rho, \tag{3.52}$$

with diffusion coefficient $D = m^2/\tau\tilde{b}$.

3.3.2.4 Telegraph equation

As discussed above, the macroscopic dynamics of randomly moving cells are well described by the diffusion equation. However, this is not the only possible macroscopic limit. The so-called hyperbolic scaling argument implies that the continuous space and time variables converge to zero with the same speed, i.e.

$$\mathbf{x} = \varepsilon \mathbf{r} \text{ and } t = \varepsilon k. \tag{3.53}$$

By expanding the time component τ of eq. (3.43) up to the second order and second order in m, one obtains the hyperbolic equation

$$\frac{\tau}{2}\partial_{tt}\rho + \varepsilon\partial_t\rho = D\nabla^2\rho, \tag{3.54}$$

where again $D = m^2/\tilde{b}\tau$. The above equation is valid only if $m \sim \tau \sim \varepsilon$ and it can be seen as a *damped wave equation*.

Equation (3.54) is an example of a *telegraph equation*, which was originally studied by Lord Kelvin in relation to signals propagating across the transatlantic cable [Goldstein 1951]. Goldstein (1951) was the first to show that the equation is also the governing equation of this special type of random walk process. Kac [Kac 1974] completed a similar analysis, and hence this general type of movement process is often termed the Goldstein-Kac model. More recently, [Shlesinger 2003] has demonstrated that a random walk with a coupled space-time memory can also be used to derive (3.54).

Let us now calculate the mean square displacement corresponding to the telegraph equation as shown in the subsec. 3.1.2 (see also [Othmer 1988]). First, we introduce a new parameter $\lambda = 1/\tau$ which represents the rate of a reorientation event, i.e. the rate of application of the reorientation operator O. Then, the cell velocity is $v = m|\mathbf{c}_i| = m/\tau = m\lambda$, where $|\mathbf{c}_i| = 1$, $\forall i \in \{1,...,\tilde{b}\}$. If we multiply eq. (3.54) by x^2, integrate over the whole domain $\mathbb{R}$ and use the fact that the density and spatial derivative of density tend to zero when $|x| \to \infty$, we have

$$\frac{d^2\langle x^2\rangle}{dt^2} + 2\lambda\frac{d\langle x^2\rangle}{dt} = 2D\lambda, \tag{3.55}$$

where $\langle x^2 \rangle = \int_\mathbb{R} x^2 \rho(x,t)dx$ is the mean square displacement of the process. If we assume that the initial conditions are $\rho(x,0) = \delta(x)$ and $\rho_x(x,0) = 0$, then the solution of the above equation is

Chapter 3. Random cell motion

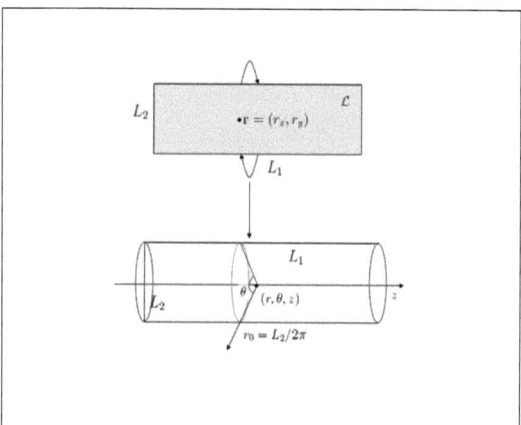

Figure 3.4: Illustration of the analogy between the evolution of a system on a lattice with periodic B.C. on axis-L_2 and the evolution of a system on a cylindrical surface ("tube"). For more details see the text (subsec. 3.3.2.5).

$$\langle x^2 \rangle = D\left\{ t - \frac{1}{2\lambda}(1 - e^{-2\lambda t}) \right\}. \tag{3.56}$$

For small $\lambda t = t/\tau \ll 1$ we expand the above formula and we have $\langle x^2 \rangle \sim \lambda D t^2$, which is characteristic of a wave process (ballistic motion); for long times $t \to \infty$, $\langle x^2 \rangle \sim Dt$, which is characteristic for the diffusion process with diffusion coefficient $D = m^2/\tau \tilde{b} = v^2/\lambda \tilde{b}$. The direct calculation of the mean square displacement $\langle x^2 \rangle$ for the diffusion eq. (3.52) coincides therefore with the long time behavior of the telegraph equation.

As we have seen above, the telegraph equation provides insight into the short-term behavior of a randomly moving cell population. Practically, ballistic motion can be seen as an accelerating regime that "speeds up" the population till the speed is proportional to $\sqrt{D}$ and then the population maintains this speed. Microscopical mechanisms that explain the ballistic regime can be considered: (i) the cell "inertia" or (ii) the short-time directional persistence or (iii) the noise on the "steering" mechanisms of the cells.

3.3.2.5 Simulations vs theory

In the following, we compare the analytical results provided in the previous sections, with LGCA simulations. Let us consider a square lattice $\mathbf{r} = (r_x, r_y) \in \mathcal{L} = L_1 \times L_2$ with periodic boundary condition (B.C.) on the L_2-axis, i.e. $n(r_x, r_y = 0, k) = n(r_x, r_y = L_2, k)$, $\forall r_y \in [0, L_2]$, $\forall k \in \mathbb{N}$ and open boundary conditions on the

3.3. Characterization of random motion

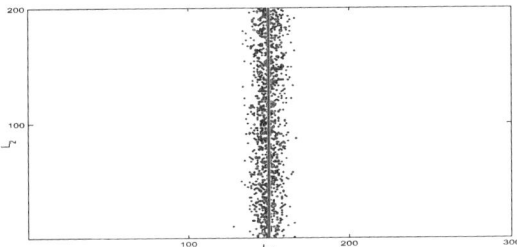

Figure 3.5: Typical simulation on a square lattice with $\tilde{b} = 6$ channels with periodic boundary condition along the L_2-axis. The colors denote the node density. In the middle the red stripe represents the initial conditions of the simulation, which is a fully occupied lattice column of cells.

L_1-axis. The initial condition of the simulations is a thin stripe of cells located along the $r_x = L_1/2$ column of the lattice. A typical simulation lasts for 100 time steps (fig. 3.5). In order to study cell motility, we reduce the two-dimensional system to one dimension by averaging the concentration profile along the L_2-axis, i.e. $n(r_x, k) = \frac{1}{|L_2|}\sum_{r_y \in |L_2|} n(\mathbf{r}, k)$. The normalized $n(r_x, k)$ is denoted as $\tilde{n}(r_x, k) = n(r_x, k)/(|L_2|\tilde{b})$. The final state of the quantity $\tilde{n}$ is approximately a Gaussian bell as shown in fig. 3.6. As before, the observable that characterizes the motility rate of the cells is the mean square displacement

$$\langle x^2 \rangle = \int_\mathbb{R} \left(x - \frac{L_1}{2}\right)^2 \rho(x,t)dx \to \frac{1}{2}\sum_{r_x \in L_1}\left(r_x - \frac{L_1}{2}\right)^2 \tilde{n}(r_x, k). \qquad (3.57)$$

It is important to show that the above simulation setting reduces the dimension of the system but it does not influence the qualitative behavior of the LGCA. In the following, we show that the diffusion coefficient is identical both in free boundary 2D simulations and in simulations with periodic B.C. on the L_2−axis.

The evolution of a system on a lattice with periodic B.C. on the L_2-axis can be seen as an evolution on a cylindrical surface with open B.C., resembling the "evolution on a tube". Therefore, the macroscopic dynamics of the system can be realized by the diffusion equation in cylindrical coordinates, as shown in fig. 3.4. In particular, eq. (3.52) in cylindrical coordinates $(x, y, z) = (r\cos\theta, r\sin\theta, z)$ yields

$$\frac{\partial}{\partial t}\rho(r,\theta,z,t) = D\left[\frac{1}{r}\frac{\partial}{\partial r}\left(r\frac{\partial}{\partial r}\right) + \frac{1}{r^2}\frac{\partial^2}{\partial \theta^2} + \frac{\partial^2}{\partial z^2}\right]\rho(r,\theta,z,t), \qquad (3.58)$$

with boundary conditions $\rho(r, \theta = 0, z, t) = \rho(r, \theta = 2\pi, z, t)$ and $\partial_\theta\rho(0) = \partial_\theta\rho(2\pi)$. As shown in fig. 3.4, the variable radial coordinate is fixed, i.e. $r = r_0 = |L_2|/2\pi$. Thus, eq. (3.58) simplifies to

Chapter 3. Random cell motion

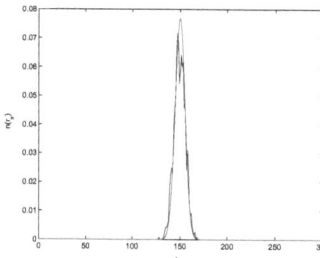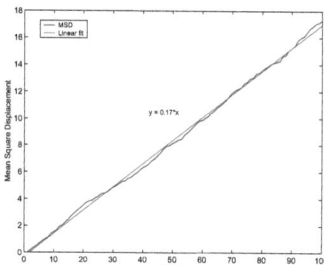

Figure 3.6: The left figure represents the averaged quantity $n(r_x, k)$ of fig. 3.5 (where $\tilde{b} = 6$). On the right, we observe the time evolution of the mean square displacement. The diffusion coefficient of the simulations, which is the slope of the line, coincides with the theoretically calculated value $D = m^2/\tilde{b}\tau = 1/6 \simeq 0.17$.

$$\rho_t = D\rho_{zz} + \frac{4\pi^2 D}{|L_2|^2}\rho_{\theta\theta}, \qquad (3.59)$$

In order to reduce the system dimension to 1D, we integrate along the angular coordinate θ and the quantity of interest is the normalized averaged cell density along the L_2-axis

$$\tilde{\rho} = \frac{1}{2\pi}\int_0^{2\pi} \rho\frac{|L_2|}{2\pi}d\theta.$$

We derive the evolution equation for $\tilde{\rho}$ by integrating eq. (3.59) over θ, previously multiplied by $|L_2|/4\pi^2$

$$\tilde{\rho}_t = D\tilde{\rho}_{zz} + \frac{2\pi D}{|L_2|^2}\int_0^{2\pi} \rho_{\theta\theta}\frac{|L_2|}{2\pi}d\theta = D\tilde{\rho}_{zz} + \frac{D}{|L_2|}\big[\rho_\theta(2\pi) - \rho_\theta(0)\big]. \qquad (3.60)$$

The last term is obviously zero due to the periodic B.C. of the system. Thus, we obtain the equation

$$\tilde{\rho}_t = D\tilde{\rho}_{zz}. \qquad (3.61)$$

The above equation shows that the "tube" simulation setup and the dimension reduction process (through averaging along the L_2-axis) to the L_1-axis does not change the diffusion coefficient of the original system. Thus, the mean square displacement will remain intact, i.e. $\langle x^2 \rangle = Dt$. In our simulations, we calculate the mean square displacement of the cell population for the square lattice with $\beta = 2$ rest channels. After 100 time steps, we observe that the mean square displacement evolves linearly in time with a slope approximately 0.17 $\simeq$ 1/6. From our analysis, we know that the slope of the mean square displacement corresponds to the

3.3. Characterization of random motion

diffusion coefficient of our cell population. Thus, the theoretical value of the diffusion coefficient $D = m^2/\tilde{b}\tau = 1/6$ coincides with the one evaluated on the basis of simulations.

However, in our simulations (fig. 3.6), we cannot observe the ballistic motion of the cells predicted in the telegraph equation. This is an expected behavior since the observation time of the ballistic regime is much smaller than the fixed time step of the simulation, i.e. $t_{ballistic} \ll \tau$. However, the following two questions arise:

- Does this mean that the telegraph equation is not the right limit of our LGCA? The answer is that the random walk dynamics on a regular lattice does not exhibit ballistic motion. However, the hyperbolic scaling serves as an analytical "magnifying glass" that allows for the extrapolation of the motion dynamics corresponding to short times $t < \tau$.

- Does this fact disqualify the LGCA as model of cell motion? The answer is negative. We claim that the short-term dynamics are resolved within the time step τ and the observed motion is just diffusion. A potential way to observe the ballistic regime is to use a LGCA with probabilistic application of the reorientation operator (O). However, we do not develop such a model in this thesis, since we are interested in the long-term dynamics of the cells.

Finally, we discuss the case where the single cell and the collective diffusion coefficient differ, i.e. $D^* \neq D$. Let's recall the extreme example at the begin of sec. 3.3. We assume a one-dimensional LGCA ($d = 1$) equipped with a lattice $\mathcal{L} \subset \mathbb{R}$ with $\tilde{b}$ channels and periodic boundary conditions. We consider the node distance $m = 1$ and the time step $\tau = 1$. The motion dynamics are obeying the random walk rule introduced in this chapter. Moreover, we assume that the lattice is fully occupied, i.e. $n(r,0) = \tilde{b}$, where $r \in \mathcal{L}$. In order to calculate the single cell diffusion coefficient, we need to use Einstein formula (3.27) and for the collective case eq. (3.8). Accordingly, the diffusion coefficient of a single-tagged cell is given by the Einstein's formula and the Green-Kubo formula (3.26)

$$D^* = \frac{\langle X^2 \rangle}{2dk} = \frac{k\langle v^2(0)\rangle_{eq}}{2k} = \frac{1}{\tilde{b}}, \quad (3.62)$$

and the mean square displacement is linearly increasing in time. This is rather expected since the tagged cell can perform a random walk within the nodes of the lattice. The things are a bit different for the collective case. Firstly, we define the normalized density $\tilde{n}(r,k) = n(r,k)/(\tilde{b}|\mathcal{L}|)$. Using eq. (3.57) we calculate the mean square displacement of the cell ensemble over the lattice

$$\langle x^2 \rangle = \frac{1}{2}\sum_{r \in L}\left(r - \frac{|\mathcal{L}|}{2}\right)^2 \tilde{n}(r,k) = \frac{(|\mathcal{L}|+2)(|\mathcal{L}|+1)}{48\tilde{b}}. \quad (3.63)$$

Here we have used the series identity $\sum_{n=1}^{k} n^2 = k(k+1)(2k+1)/6$ and the fact that the node density remains invariant, i.e. the nodes remain all the time fully occupied $n(r,k) = \tilde{b}$, $\forall r \in \mathcal{L} \wedge \forall k \in \mathbb{N}$. Since the nodes of the lattice remain full, we do not expect any expansion of the cell ensemble for any given time, i.e. the mean square displacement is constant. Using eq. (3.8), the collective diffusion coefficient yields

$$D = \frac{1}{2}\frac{d\langle x^2 \rangle}{dt} = 0. \qquad (3.64)$$

Therefore, we observe that for a specific choice of initial conditions our random motion process provides different individual and collective diffusion coefficients, i.e. $D \neq D^*$.

3.4 Summary

In this chapter, we have introduced a LGCA model for random cell motion. The most important characteristic of moving cells is their motility rate. We estimated the motility rates of two distinct cases: (i) a single cell motion and of (ii) the dispersion of a cell ensemble. In the first case, we have derived the Green-Kubo formula for LGCA and we have calculated the diffusion rate of a single moving cell. In the second case, we have derived the macroscopic dynamics of a diffusive cell population by three different scaling methods: the Chapman-Enskog method, the CML method and the Fourier space method. Interestingly, all three methods provide the same macroscopic dynamics, given by the diffusion equation. Due to the simplicity of the microscopic random walk the scaling methods perform equally well. However, in the next chapter we identify some of the limitations of these methods when one wants to integrate more complicated processes. Additionally, we have derived a different macroscopic limit, i.e. the telegraph equation which allows for the description of motion for very short times. Finally, we found out that the simulation results agree with the analytical theory.

CHAPTER 4

Growth processes

Contents

4.1	Introduction	41
4.2	A LGCA growth process	43
	4.2.1 LGCA dynamics	44
	4.2.2 Microdynamical equations	45
4.3	Mean-field analysis	46
	4.3.1 Well-stirred system	46
	4.3.2 Spatially distributed system	48
4.4	Mean-field approximations and macroscopic behavior	55
4.5	Summary	57

4.1 Introduction

In this chapter, we introduce some aspects of the mathematical modeling of growth processes. Paradigms of growth processes can be found in almost any scientific field, such as physics, ecology, sociology, epidemiology, biology etc. In this thesis, we are interested in the growth of cell populations and in particular in the growth of tumors (especially discussed in a following chapter).

Mathematical modeling of population growth processes has a long history. Detailed reviews of various growth models with applications can be found in [Doucet 1992] and in [Banks 1994]. The first mathematical model of a population growth was due to Leonardo Pisano (Fibonacci) in 1202. More sophisticated growth models have been developed in the 18–19th century by the Englishmen *T. Malthus* (1798), *B. Gompertz* (1825) and the Belgian mathematician *P. Verhulst* (1838). All three have studied population growth models. Let $\bar{\rho}(t)$ denote the total number of individuals in a population at time t. Malthus assumed that the *per capita growth rate* of a population is constant, i.e.

$$\gamma = \frac{\bar{\rho}'(t)}{\bar{\rho}(t)}, \tag{4.1}$$

where $\bar{\rho}'(t) = d\bar{\rho}(t)/dt$. In other words, the mean growth rate of each individual that belongs to the population remains constant. The solution of eq. (4.1) is

$$\bar{\rho}(t) = \bar{\rho}(0)e^{\gamma t}. \quad (4.2)$$

The above formula is known as *exponential growth* law. Exponential growth captures the short-term behavior of a "doubling" process, but generally remains an unrealistic model for long times, since it implies unlimited growth. A more realistic approach has been developed by Verhulst who was taking into account the influence of limited resources or "crowding effects". In particular, Verhulst proposed that the per capita growth rate should linearly decrease with the size of the population, and modeled this property by

$$\frac{\bar{\rho}'(t)}{\bar{\rho}(t)} = r\left(1 - \frac{\bar{\rho}(t)}{K}\right), \quad (4.3)$$

where r is the reproduction rate and K is the carrying capacity of the population, i.e. the maximum population size that can be supported by the existing resources. Verhust's model (4.3) is commonly known as the *logistic growth model*. A third growth law frequently met in the literature is the *Gompertzian law*. Gompertz proposed that the mean growth rate of the individuals should be exponentially decaying with time

$$\frac{\bar{\rho}'(t)}{\bar{\rho}(t)} = re^{-at}, \quad (4.4)$$

where a, r are parameters of the model. The solution of the above equation is

$$\bar{\rho}(t) = \bar{\rho}(0)e^{\frac{r}{a}(1-e^{-at})} = Ke^{-\frac{r}{a}e^{-at}}, \quad (4.5)$$

where $K = \bar{\rho}(0)\exp(r/a)$ and $\bar{\rho}(0)$ the initial population. Using the above relations, we observe that there is a connection between the logistic growth model and the Gompertzian law, since eq. (4.4) can be rewritten as

$$\frac{\bar{\rho}'(t)}{\bar{\rho}(t)} = a\big(\ln(K) - \ln(\bar{\rho}(t))\big), \quad (4.6)$$

This model has been widely used to describe the growth of solid tumors in animal experiments [Casey 1934].

The above models do not take into account the dispersal of individuals in space. Partial differential equation (PDE) models have been employed to describe the spatio-temporal evolution of population growth processes. The most famous model is the non-dimensional *Fisher-Kolmogorov* model

$$\partial_t \rho(\mathbf{x}, t) = \nabla^2 \rho(\mathbf{x}, t) + \rho(\mathbf{x}, t)(1 - \rho(\mathbf{x}, t)), \quad (4.7)$$

where $\rho(\mathbf{x}, t)$ denotes the local density of individuals. The Fisher-Kolmogorov model describes a diffusive population with logistic growth dynamics, which is a particular kind of reaction-diffusion equations (RDE). In the literature, one can find numerous continuous spatio-temporal growth models in the context of RDEs, where the diffusion term refers to cell migration and the reaction term models the kind of growth (or annihilation) process.

Apart from the above-mentioned continuous models, there is a vast literature on discrete models in the form of difference equations. The most basic discrete model is the logistic map, which is the discrete representation of Verhulst's model (4.3), i.e.

$$\rho_{k+1} = \bar{r}\rho_k\left(1 - \frac{\rho_k}{K}\right), \quad (4.8)$$

where ρ_k is the number of individuals at the discrete time $k \in \mathbb{N}$ and $\bar{r}$ the reproduction rate of an individual within a time step. Note, that the discrete logistic map may exhibit qualitatively different behavior from the continuous one for specific parameter values. When the dynamics become probabilistic the growth of a population can be described by a special class of stochastic Markov processes, the so-called birth/death processes [Gardiner 1990, Norris 1997]. In particular, the evolution of the quantity ρ_k now depends on probabilities of creation or annihilation of one or more individuals.

The consideration of space in discrete growth processes led to the development of discrete spatio-temporal models, such as cellular automata. In this thesis the focus is on LGCA, which are able to model spatially distributed growth processes of cell populations (from now on in this chapter, we will refer to cells instead of individuals). Here, we provide an example of a logistic growth process modeled by a LGCA, proposed in [Boerlijst 2006]. We will directly compare the analytical predictions coming from the mean-field approximation of the growth process, under two different assumptions, to the stochastic LGCA simulations. First, under the assumption of a well-stirred system, we derive the temporal mean-field equation of the LGCA. Then, assuming a spatially distributed system and based on the spatio-temporal mean-field description of the microscopic growth process, we calculate the corresponding macroscopic PDE, which is a reaction-diffusion equation of Fisher-Kolmogorov type, by means of (i) the Chapman-Enskog, (ii) the Coupled Map Lattice (CML) and (iii) the Fourier space method. Simulation results of the LGCA are found to be in good agreement with the mean-field PDE, in contrast to the purely temporal mean-field approximation. From this mean-field PDE we derive characteristic macroscopic observables of biological growth, such as total number of cells and per capita growth rate, and we reveal their dependence on the microscopic growth and transport parameters. Finally, we discuss the characteristic properties of the different scaling methods that allow for the derivation of the macroscopic dynamics.

4.2 A LGCA growth process

We consider a lattice-gas cellular automaton [Deutsch 2005] defined on a two-dimensional regular, square lattice $\mathcal{L} = L_1 \times L_2 \subset \mathbb{Z}^2$, where L_1 and L_2 are the lattice dimensions. Cells move on the discrete lattice with discrete velocities, i.e.

they hop at discrete time steps from a given node to a neighboring one, as determined by the *single particle speed* m. The set of velocities for the square lattice as considered here, is represented by the two-dimensional channel velocity vectors $\mathbf{c_1} = \begin{pmatrix} 1 \\ 0 \end{pmatrix}$, $\mathbf{c_2} = \begin{pmatrix} 0 \\ 1 \end{pmatrix}$, $\mathbf{c_3} = \begin{pmatrix} -1 \\ 0 \end{pmatrix}$, $\mathbf{c_4} = \begin{pmatrix} 0 \\ -1 \end{pmatrix}$, $\mathbf{c_5} = \begin{pmatrix} 0 \\ 0 \end{pmatrix}$, (see fig. 2.2). In each of these channels, we consider an exclusion principle, i.e. we allow at most one cell per channel. We denote by $\tilde{b} = b + \beta$ the total number of channels per node which can be occupied simultaneously, where β is the number of channels with zero velocity, the so-called rest channels. We represent the channel occupancy by a Boolean random variable called *occupation number* $\eta_i(\mathbf{r}, k) \in \{0, 1\}$, where $i = 1, ..., \tilde{b}$, where $\mathbf{r} = (r_x, r_y) \in \mathcal{L} \subset \mathbb{Z}^2$ denotes the spatial variable and $k \in \mathbb{N}$ the discrete time variable.

4.2.1 LGCA dynamics

In our model, cell dynamics are identified by the LGCA rules. Automaton dynamics arise from the repetition of three rules (operators): Propagation (P), reorientation (O) and growth (R). In particular, the reorientation and the propagation operators are related to the cell motion (which have been discussed in Chapter 3) and the growth operator controls the change of the local number of cells on a node and is the focus of this chapter. In the following, we present in detail the LGCA operators related to our growth process.

4.2.1.1 Propagation (P)

As discussed in the previous chapters, the process of cell movement in the medium is modeled by the propagation step. The propagation step is deterministic and it is governed by an operator P. By the application of P all cells are transported simultaneously to nodes in the direction of their velocity, i.e. a cell residing in channel $(\mathbf{r}, \mathbf{c}_i)$ at time k is moved to a neighboring channel $(\mathbf{r} + m\mathbf{c}_i, \mathbf{c}_i)$ during one time step (fig. 2.4). Here $m \in \mathbb{N}$ determines the speed and $m\mathbf{c}_i$ is the translocation of the cell. The cells residing on the rest channel do not move since they have zero velocity. In terms of occupation numbers, the state of a channel $(\mathbf{r} + m\mathbf{c}_i, \mathbf{c}_i)$ after propagation becomes:

$$\eta_i(\mathbf{r} + m\mathbf{c}_i, k + \tau) = \eta_i^P(\mathbf{r} + m\mathbf{c}_i, k) = \eta_i(\mathbf{r}, k),$$

where $\tau \in \mathbb{N}$ is the automaton's time-step. We note that this operator is mass and momentum conserving.

4.2.1.2 Reorientation (O)

The reorientation operator is responsible for the redistribution of cells within the velocity channels of a node, providing a new node velocity distribution as discussed in the previous chapter. Here, we assume that the cells are just random walkers. A possible choice for the corresponding transition probabilities is

4.2. A LGCA growth process

$$\mathbb{P}(\boldsymbol{\eta} \to \boldsymbol{\eta}^{\mathrm{O}})(\mathbf{r}, k) = \frac{1}{Z}\delta\big(n(\mathbf{r}, k), n^{\mathrm{O}}(\mathbf{r}, k)\big), \tag{4.9}$$

where $Z = \sum_{\boldsymbol{\eta}^{\mathrm{O}}(r,k)} \delta\big(n(\mathbf{r}, k), n^{\mathrm{O}}(\mathbf{r}, k)\big)$ is a normalization factor. The Kronecker δ assumes the mass conservation of this operator. The impact of this rule on collective cell motion has been analyzed in Chapter 3.

4.2.1.3 Growth (R)

In our model, cells are allowed to proliferate if there is sufficient space. In particular, we assume a microscopic volume exclusion growth dynamics, where the maximum capacity is defined by the node capacity $\tilde{b}$. The effect of local volume exclusion on growth is also known as *carrying capacity-limited* or *contact-inhibited* growth. For the creation of a new cell on a node at least one free channel is required. This condition can be formulated in the following way:

$$\mathcal{R}_i(\mathbf{r}, k) = \xi_i(\mathbf{r}, k)(1 - \eta_i(\mathbf{r}, k)), \tag{4.10}$$

where $\xi_i(\mathbf{r}, k)$'s are random Boolean variables, with $\sum_{i=1}^{\tilde{b}} \xi_i(\mathbf{r}, k) = 1$, and the corresponding probabilities are:

$$\mathbb{P}(\xi_i(\mathbf{r}, k) = 1) = \mathbb{P}\big(\{\eta_i^{\mathrm{R}}(\mathbf{r}, k) = 1\}, \{n(\mathbf{r}, k) \neq 0\}\big) \in \{0, 1\}, \tag{4.11}$$

which is the probability that a newly created cell occupies a channel i and that there exists at least one cell on node $\mathbf{r}$. For analytical calculations, we assume the independency of these two events. Moreover, we define $r_m = \mathbb{P}\big(\{\eta_i^{\mathrm{R}}(\mathbf{r}, k) = 1\}\big)$ as the probability of a new cell occupying channel i. Altogether we have

$$\mathbb{P}(\xi_i(\mathbf{r}, k) = 1) = \mathbb{P}\big(\{\eta_i^{\mathrm{R}}(\mathbf{r}, k) = 1\}\big)\mathbb{P}\big(\{n(\mathbf{r}, k) \neq 0\}\big) = r_m \frac{n(\mathbf{r}, k)}{\tilde{b}}. \tag{4.12}$$

4.2.2 Microdynamical equations

The above defined dynamics is fully specified by the following microdynamical equations:

$$\eta_i^{\mathrm{R}}(\mathbf{r}, k) = \eta_i(\mathbf{r}, k) + \mathcal{R}_i(\mathbf{r}, k), \tag{4.13}$$

$$\eta_i(\mathbf{r} + m\mathbf{c}_i, k + \tau) = \sum_{j=1}^{\tilde{b}} \mu_j(\mathbf{r}, k)\eta_j^{\mathrm{R}}(\mathbf{r}, k). \tag{4.14}$$

The first eq. (4.13) refers to the application of the *growth* operator (R), which assigns a new occupation number for a given channel through a stochastic growth process. The second equation (4.14) refers to the *redistribution* of cells on the velocity channels and the *propagation* to the neighboring nodes, corresponding to the random walk as introduced in the previous chapter.

The $\mu_j(\mathbf{r}, k) \in \{0, 1\}$ are Boolean random variables which select only one of the $\tilde{b}$ terms of the rhs of eq. (4.14). Therefore, they should satisfy the relation $\sum_{j=1}^{\tilde{b}} \mu_j = 1$. As stated above, we implement the random walk as a simple reshuffling of the cells within the node channels that leads to the probability of choosing a channel: $\langle \mu_j \rangle = 1/\tilde{b}$, for $j = 1, ..., \tilde{b}$ (see Chapter 2). The terms $\mathcal{R}_i(\mathbf{r}, k) \in \{0, 1\}$, for $i = 0, ... \tilde{b}$ eq. (4.10) represent birth processes, i.e. creation of cells in channel i defined by the growth rule, which are applied to each channel independently.

4.3 Mean-field analysis

This section introduces the mean-field (MF) analysis of the above defined growth LGCA. The main idea of the mean-field approximation is to replace the description of many-cell interactions by a single cell description based on an average or effective interaction. Thereby, any multi-cellular problem can be replaced by an effective problem, that can be stated in the form of a macroscopic description such as an ordinary (ODE) or a partial (PDE) differential equation.

We introduce the mean-field analyisis under the assumption of a well-stirred case (for which the diffusion coefficient diverges) and a spatially distributed case, i.e. a finite diffusion strength ($< \infty$).

4.3.1 Well-stirred system

Here, we derive a mean-field approximation of our LGCA under the assumption of a well-stirred case [Kapral 1997]. In automaton terms, within one time step $(k, k+\tau)$ the transport operator (4.14), which randomly reshuffles the cells on the velocity channels and propagates them to the neighboring nodes, is repeatedly applied till the system is homogenized and relaxes to a single binomial distribution over the lattice, i.e. $(P \circ O)^{l_D}$, where $l_D \to \infty$ [Wu 1994]. Phenomenologically, the well-stirred system corresponds to a divergent diffusion coefficient, i.e. $D \to \infty$. That means that the characteristic time of growth is much slower than the characteristic time related to cell motion (diffusion time). Applying the mean-field approximation, i.e. neglecting all spatial correlations, we define:

$$P_{binom}(n(\mathbf{r}), \rho) = \frac{\tilde{b}!}{(\tilde{b} - n(\mathbf{r}))! n(\mathbf{r})!} \left(\frac{\rho}{\tilde{b}}\right)^{n(\mathbf{r})} \left(1 - \frac{\rho}{\tilde{b}}\right)^{\tilde{b} - n(\mathbf{r})}, \forall \mathbf{r} \in \mathcal{L}, \quad (4.15)$$

where ρ is the average node density of the lattice, i.e. $\rho = \sum_{\mathbf{r} \in \mathcal{L}} n(\mathbf{r})/|\mathcal{L}|$. We drop the temporal k and the spatial argument $\mathbf{r}$ since the well-stirred system assumes a spatially homogeneous node density distribution. The joint probability of the set of node configurations $\{\boldsymbol{\eta}(\mathbf{r})\}_{\mathbf{r} \in \mathcal{K}}$, where $\mathcal{K} = \{\mathbf{r} \in \mathcal{L} | n(\mathbf{r}) \neq 0\} \subseteq \mathcal{L}$ is the set of occupied nodes of the lattice $\mathcal{L}$, is:

$$\mathbb{P}(\{\boldsymbol{\eta}(\mathbf{r})\}_{\mathbf{r} \in \mathcal{K}}) = \prod_{\mathbf{r} \in \mathcal{K}} P_{binom}(n(\mathbf{r}), \rho). \quad (4.16)$$

4.3. Mean-field analysis

By the application of the reaction step, the node density may change to $n(\mathbf{r}) \to n(\mathbf{r}) + 1$ with the transition probability $p^+(n(\mathbf{r}))$. Let α_+ denote the nodes that gain a cell, where this event is realized by the probability:

$$\mathbb{P}(\alpha_+) = \sum_{\{\boldsymbol{\eta}(\mathbf{r})\}_{\mathbf{r}\in\mathcal{K}}} \mathbb{P}(\alpha_+|\{\boldsymbol{\eta}(\mathbf{r})\})\mathbb{P}(\{\boldsymbol{\eta}(\mathbf{r})\}). \tag{4.17}$$

Here $\mathbb{P}(\alpha_+|\{\boldsymbol{\eta}(\mathbf{r})\})$ is the conditional probability for the transition $n(\mathbf{r}) \to n(\mathbf{r})+1$ from a given configuration $\{\boldsymbol{\eta}(\mathbf{r})\}_{\mathbf{r}\in\mathcal{K}}$. Therefore,

$$\mathbb{P}(\alpha_+|\{\boldsymbol{\eta}(\mathbf{r})\}) = \prod_{\mathbf{r}\in[1,\alpha_+]} p^+\bigl(n(\mathbf{r})\bigr) \prod_{\mathbf{r}\in[\alpha_++1,|\mathcal{K}|]} \bigl(1 - p^+(n(\mathbf{r}))\bigr). \tag{4.18}$$

Summing over all possible configurations $\{\boldsymbol{\eta}(\mathbf{r})\}_{\mathbf{r}\in\mathcal{K}}$ in (4.17), we obtain the following binomial distribution:

$$\mathbb{P}(\alpha_+) = \frac{|\mathcal{K}|!}{\alpha_+!(|\mathcal{K}|-\alpha_+)!}(P^+)^{\alpha_+}(1-P^+)^{|\mathcal{K}|-\alpha_+}, \tag{4.19}$$

where the prefactor takes into account the number of all possible choices of assigning the α_+ transitions $n \to n+1$ to possible positions on the lattice. The probabilities $P^+ = \sum_n p^+(n) P_{binom}(n, \rho)$ are the averages of the transition probabilities $p^+(n)$ over the binomial distribution on the lattice $\mathcal{L}$.

Therefore, the expectation value of the net change in the node density during one time step becomes:

$$\langle \Delta\rho \rangle_{MF} = \rho(k+\tau) - \rho(k) = \frac{1}{|\mathcal{K}|}\sum_{\alpha_+\leq|\mathcal{K}|} \alpha_+ \mathbb{P}(\alpha_+) = P^+ = F(\rho). \tag{4.20}$$

where $F(\rho)$ is the mean-field growth law for a single node. Following the definition of the automaton's reaction rule (4.10), the mean-field growth term is:

$$F(\rho) = r_m \rho (1 - \frac{\rho}{\tilde{b}}). \tag{4.21}$$

The transition of a microscopical process to a macroscopic description requires a *temporal scaling* relation between the macroscopic and microscopic variables. We assume a small parameter $\varepsilon \ll 1$ that scales the time variable t:

$$t = \varepsilon k. \tag{4.22}$$

Using the Taylor expansion $\rho(k+\tau) = \rho(k) + \varepsilon\tau\partial_t\rho(t)\big|_{t=k}$ and rewriting equation (4.20) in continuous variables:

$$\partial_t \rho = \frac{r_m}{\varepsilon\tau}\rho(1 - \frac{\rho}{\tilde{b}}), \tag{4.23}$$

which is the *logistic growth* equation. Note that $0 < \varepsilon\tau \ll 1$. Since the ratio $r_m/\varepsilon\tau < \infty$ should be finite for $\varepsilon \to 0$, the growth rate should be scaled as $r_m \propto \varepsilon\tau \ll 1$, which means that the equation (4.23) is valid for small growth rates. Therefore, we have shown that our growth process under the mean-field and the well-stirred system assumptions behaves macroscopically as the logistic equation (4.23).

4.3.2 Spatially distributed system

In the previous section, we have derived the mean-field approximation under the well-stirred system assumption. Here, we demonstrate how to derive a spatio-temporal mean-field approximation of the growth LGCA model for a spatially distributed system. As introduced in the previous chapter, the average change of the occupation number of channel i, where $i = 1, ..., \tilde{b}$, is:

$$f_i(\mathbf{r} + m\mathbf{c}_i, k + \tau) - f_i(\mathbf{r}, k) = \langle \eta_i(\mathbf{r} + m\mathbf{c}_i, k + \tau) - \eta_i(\mathbf{r}, k) \rangle_{MF}. \quad (4.24)$$

Under the mean-field approximation (MF), where all spatial correlations are neglected, the distribution functions are factorized. Combining equations (4.13), (4.14) and (4.24), we obtain the LBE:

$$f_i(\mathbf{r} + m\mathbf{c}_i, k + \tau) - f_i(\mathbf{r}, k) = \sum_{j=1}^{\tilde{b}} \mathbf{\Omega}_{ij} f_j(\mathbf{r}, k) + \sum_{j=1}^{\tilde{b}} (\delta_{ij} + \mathbf{\Omega}_{ij}) \bar{\mathcal{R}}_i(\mathbf{r}, k), \quad (4.25)$$

where the matrix $\mathbf{\Omega}_{ij} = 1/\tilde{b} - \delta_{ij}$ is the transition matrix of the underlying shuffling process (reorientation operator). Moreover, we assume that the mean-field growth term is independent of the cell direction, i.e. $\bar{\mathcal{R}}_i = F(\rho)/\tilde{b}$, where $F(\rho)$ is the mean-field growth term for a single node. The mean-field growth term is:

$$\bar{\mathcal{R}}_i(\mathbf{r}, k) = r_m f_i(\mathbf{r}, k) \big(1 - f_i(\mathbf{r}, k)\big), \quad (4.26)$$

where r_m is the probability of the birth of a new cell.

4.3.2.1 Chapman-Enskog method

In order to derive the macroscopic dynamics, we use the Chapman-Enskog methodology. As in the previous chapter (subsec. 3.3.2.1), we assume the parabolic (or diffusive) spatio-temporal scaling

$$\mathbf{x} = \varepsilon \mathbf{r} \text{ and } t = \varepsilon^2 k, \quad (4.27)$$

where $(\mathbf{x}, t)$ are the continuous variables as $\varepsilon \to 0$. Using the spatio-temporal scaling relation eq. (4.27) and replacing the first part of eq. (4.25) by its Taylor expansion, we obtain eq. (3.32).

Now, we assume an asymptotic solution of the single particle distribution f_i in terms of the parameter ε:

$$f_i = f_i^{(0)} + \varepsilon f_i^{(1)} + \varepsilon^2 f_i^{(2)} + \mathcal{O}(\varepsilon^3). \quad (4.28)$$

An important aspect is the scaling of the growth term. We assume that the birth of cells is taking place at a much slower time scale than the motion. The idea is that growth can be considered as a perturbation of cell motion. That means that

4.3. Mean-field analysis

the dominant process is cell diffusion (as it is shown below). The growth rate is assumed to be scaled according to the macroscopic time scaling, i.e.

$$\mathcal{R}_i \to \varepsilon^2 \tilde{\mathcal{R}}_i. \tag{4.29}$$

The eq. 4.29 provides that the macroscopic rate shall be scaled as $r_m = \varepsilon^2 \tilde{r}_m \ll 1$, where $\tilde{r}_m = \mathcal{O}(1)$. Therefore, our approach is valid only for very low growth rates.

The macroscopic quantity of interest is the cell density $\rho = \sum_i f_i^{(0)}$, with the assumption $\sum_i f_i^{(l)} = 0$, if $l \geq 1$. This means that only the equilibrium solution $f_i^{(0)}$ contributes to the mass of the system.

The next step is to insert (4.27) and (3.32) into (4.25) and to collect the terms of equal ε order:

$$\mathcal{O}(\varepsilon^0) \; : \; \sum_j \Omega_{ij} f_j^{(0)} = 0, \tag{4.30}$$

$$\mathcal{O}(\varepsilon^1) \; : \; m(\mathbf{c}_i \cdot \nabla) f_j^{(0)} = \sum_j \Omega_{ij} f_j^{(1)}, \tag{4.31}$$

$$\mathcal{O}(\varepsilon^2) \; : \; \tau \partial_t f_i^{(0)} + m(\mathbf{c}_i \cdot \nabla) f_i^{(1)} + \frac{m^2}{2} (\mathbf{c}_i \cdot \nabla)^2 f_i^{(0)} = \sum_j \Omega_{ij} f_j^{(2)} + \tilde{R}_i. \tag{4.32}$$

The solutions of the above equations (4.30), (4.31), (4.32) are based on the properties of the transition matrix Ω [Chopard 1998] and the results are:

$$f_j^{(0)} = \frac{\rho}{\tilde{b}}, \tag{4.33}$$

$$f_j^{(1)} = -\frac{m}{\tilde{b}} (\mathbf{c}_i \cdot \nabla) \rho. \tag{4.34}$$

To obtain a macroscopic equation, we sum equation (4.32) over i, which yields:

$$\tau \partial_t \sum_i f_i^{(0)} + m \sum_i (\mathbf{c}_i \cdot \nabla) f_i^{(1)} + \frac{m^2}{2} \sum_i (\mathbf{c}_i \cdot \nabla)^2 f_i^{(0)} = \sum_{ij} \Omega_{ij} f_j^{(2)} + \sum_i \tilde{R}_i. \tag{4.35}$$

Performing the above summation (using the property of the transition matrix $\sum_{ij} \Omega_{ij} = 0$ and the property of the lattice tensor[1] $\sum_i c_{i\alpha} c_{i\beta} = d\delta_{\alpha\beta}$, where d is the dimension of the system), yields the following reaction-diffusion equation:

$$\partial_t \rho = \frac{m^2}{\tilde{b}\tau} \nabla^2 \rho + \frac{1}{\tau} F(\rho), \tag{4.36}$$

where $F(\rho) = \tilde{b}\tilde{R}_i$ is the macroscopic growth law, i.e.

$$F(\rho) = \tilde{r}_m \rho (1 - \frac{\rho}{\tilde{b}}). \tag{4.37}$$

---

[1] The automaton's spatial and velocity discretizations are related to the definition of the lattice tensor. The velocity vectors $\mathbf{c}_i$ yield an orthogonal basis for the construction of the lattice tensors. The lattice tensors contain all the isotropy and symmetry properties of the automaton's space $\mathcal{L}$ [Rothman 1994].

By non-dimensionalizing the above equation, we can readily obtain the well-known *Fisher-Kolmogorov (FK) equation* (4.7).

Moreover, the Chapman-Enskog procedure allows us to derive an asymptotic solution for $f_i(\mathbf{r},k)$ (see appendix), i.e.

$$f_i(\mathbf{r},k) = \frac{\rho(\mathbf{r},k)}{\tilde{b}} - \varepsilon \frac{m}{\tilde{b}}(\mathbf{c}_i \cdot \nabla)\rho(\mathbf{r},k) + \varepsilon^2 \frac{m^2}{2\tilde{b}^2}\left[\tilde{b}(\mathbf{c}_i \cdot \nabla)^2 - 2\nabla^2\right]\rho(\mathbf{r},k). \quad (4.38)$$

Intuitively, we can understand the above expression as a perturbation of the homogeneous equilibrium solution $\rho(\mathbf{r},k)/\tilde{b}$.

4.3.2.2 Limit of the CML equation

Initially, we average the system of microdynamical equations (4.13-4.14) and we sum over the $\tilde{b}$ channels:

$$\rho^R(\mathbf{r},k) = \rho(\mathbf{r},k) + F(\rho(\mathbf{r},k)), \quad (4.39)$$

$$\rho(\mathbf{r},k+\tau) = \frac{1}{\tilde{b}}\sum_{j=1}^{\tilde{b}} \rho^R(\mathbf{r} - m\mathbf{c}_j, k), \quad (4.40)$$

where $F(\rho) = r_m \rho(1 - \frac{\rho}{b})$ is the mean-field growth term as defined in eq. (4.37). The first equation accounts for the average change in the node density per time step. The new population of cells $\rho^R(\mathbf{r},k)$ is redistributed to the neighboring nodes according to eq. (4.40). We can rewrite eq. (4.40) as

$$\rho(\mathbf{r},k+\tau) = (1-\mu)\rho^R(\mathbf{r},k) + \frac{\mu}{b}\sum_{i=1}^{b}\rho^R(\mathbf{r}-m\mathbf{c}_i,k), \quad (4.41)$$

where $\mu = \frac{b}{\tilde{b}}$ is the fraction of cells that remain at the current node. The above CML implies that the new node density is given by the average density of the local node neighborhood.

Combining equations (4.39-4.40), we can write down the following integro-difference equation:

$$\rho(\mathbf{r},k+\tau) = \int_{\mathcal{L}} \left[\rho(\mathbf{r}-m\mathbf{c}_i,k) + F(\rho(\mathbf{r}-m\mathbf{c}_i,k))\right]\phi(m)dm, \quad (4.42)$$

where $\phi(+m) = \phi(-m)$ is an isotropic (symmetric) redistribution kernel of the process (alternatively it represents the jump rate of a cell of length m). In our case, the discrete redistribution kernel is defined by:

$$\phi(m) = (1-\mu)\delta(\mathbf{r}) + \frac{\mu}{b}\sum_{i=1}^{b}\delta(\mathbf{r}-m\mathbf{c}_i) = \frac{\beta}{\tilde{b}}\delta(\mathbf{r}) + \frac{1}{\tilde{b}}\sum_{i=1}^{b}\delta(\mathbf{r}-m\mathbf{c}_i). \quad (4.43)$$

Since these kernels should be mass conserving, $\int_{\mathcal{L}} \phi(m)\,dm = 1$. Note that eq. (4.42) is valid for *any mitotic rate* $r_m \in [0,1]$.

4.3. Mean-field analysis

A closer look on the reaction term of eq. (4.42) yields:

$$\int_{\mathcal{L}} F\big(\rho(\mathbf{r} - m\mathbf{c}_i, k)\big)\phi(m)dm = \frac{\beta}{b}F\big(\rho(\mathbf{r})\big) + \frac{1}{b}\sum_{i=1}^{b} F\big(\rho(\mathbf{r} - m\mathbf{c}_i)\big). \qquad (4.44)$$

After some algebra the above equation reads

$$\int_{\mathcal{L}} F\big(\rho(\mathbf{r} - m\mathbf{c}_i, k)\big)\phi(m)dm = F\big(\rho(\mathbf{r})\big) + \frac{b}{b}\Delta_{\mathbf{r}} F\big(\rho(\mathbf{r} - m\mathbf{c}_i, k)\big), \qquad (4.45)$$

where $\Delta_{\mathbf{r}} = \big[\sum_{i=1}^{b} F\big(\rho(\mathbf{r} - m\mathbf{c}_i)\big) - bF\big(\rho(\mathbf{r})\big)\big]/b$ is the discrete spatial Laplacian. Expanding the integro-difference equation (4.42) into a Taylor series and using eq. (4.45), we obtain the following differential equation:

$$\partial_t \rho = \frac{m^2}{b\tau}\nabla^2 \rho + \frac{1}{\tau}F(\rho) + \frac{bm^2}{b\tau}\nabla^2 F(\rho). \qquad (4.46)$$

Since we are interested in the long-term dynamics of the system, we assume that the $(\mathbf{x}, t)$ variables converge to zero according to the parabolic scaling (4.27). Then eq. (4.46) becomes

$$\partial_t \rho = D\nabla^2 \rho + \varepsilon^2 Dr_m b(1 - 2\rho)\nabla^2 \rho - 2\varepsilon^2 Dr_m b(\nabla \rho)^2 + \frac{r_m}{\tau}\rho(1 - \rho), \qquad (4.47)$$

where $D = m^2/b\tau$. When the scaling parameter $\varepsilon \to 0$ and the reaction rate is scaled as $r_m = \varepsilon^2 \tilde{r}_m$ (see previous section), the above equation coincides with (4.36) derived by means of the Chapman-Enskog method, i.e.

$$\partial_t \rho = \frac{m^2}{b\tau}\nabla^2 \rho + \frac{\tilde{r}_m}{\tau}\rho(1 - \frac{\rho}{b}), \qquad (4.48)$$

Note that for another scaling argument the limiting macroscopic equation (4.46) can be highly nonlinear, due to the differentiation of the growth terms.

4.3.2.3 Fourier space method

To apply this method, we need to linearize the LBE (4.25) around a fixed point of our growth LGCA. Therefore, we need to define the single particle distribution fluctuations around a fixed point $\bar{f}$

$$\delta f_i(\mathbf{r}, k) = f_i(\mathbf{r}, k) - \bar{f}. \qquad (4.49)$$

The fixed points of the LBE (4.25) are principally determined by the zeros[2] of the growth term $\bar{\mathcal{R}}_i$, i.e. $(\bar{f}_1, \bar{f}_2) = (0, \tilde{b})$.

---
[2] The zeros of the growth term are defined as the elements of the kernel of the function $\bar{\mathcal{R}}_i$, i.e. $\ker \bar{\mathcal{R}}_i = \{f_i \in [0, 1] | \bar{\mathcal{R}}_i(f_i) = 0\}$.

The LBEs (4.25) for each channel define a $\tilde{b}$-dimensional system of difference equations with respect to the vector of variables $\mathbf{f} = (f_1,...,f_{\tilde{b}})(\mathbf{r},k)$. Thus it can be written in matrix form as

$$\mathbf{f}(\mathbf{r}+m\mathbf{c}_i, k+\tau) = \mathbf{B}\big(\mathbf{f} + \bar{\mathbf{R}}(\mathbf{f})\big)(\mathbf{r},k), \qquad (4.50)$$

where $\mathbf{B} = \mathbf{I} + \mathbf{\Omega}$ is a matrix with dimensions $\tilde{b} \times \tilde{b}$ and $\mathbf{\Omega}_{ij} = 1/\tilde{b} - \delta_{ij}$, with $i,j = 1,...,\tilde{b}$. The vector $\bar{\mathbf{R}}(\mathbf{f}) = (\bar{\mathcal{R}}_i)_{i=1,...,\tilde{b}}$ denotes the mean-field growth term (4.37) for each channel.

Now, we linearize the nonlinear LBE (4.50) around a steady state $\bar{f}$ of the system. In particular, in eq. (4.50) the only nonlinear term is the vector $\bar{\mathbf{R}}(\mathbf{f})$. Thus, we define the Jacobian matrix $\mathbf{J}$ of the growth term

$$\mathbf{J}_{ij} = \left.\frac{\partial \bar{\mathcal{R}}_i}{\partial f_j}\right|_{\bar{f}}(\mathbf{r},k), \quad i,j = 1,...,\tilde{b}. \qquad (4.51)$$

Now the linearized LBE is

$$\mathbf{f}(\mathbf{r}+m\mathbf{c}_i, k+\tau) = \mathbf{\Gamma}\mathbf{f}(\mathbf{r},k) = (\mathbf{B}+\mathbf{J})\mathbf{f}(\mathbf{r},k), \qquad (4.52)$$

where $\mathbf{\Gamma} = \mathbf{B} + \mathbf{J}$. Next, we choose to linearize around the trivial fixed point $\bar{f} = \bar{f}\mathbf{1} = 0$. This choice of steady state is not arbitrary since the zero density nodes are of particular interest in invasive processes. Typically invasive processes are described by fronts and the behavior of the tip of a front is characterized by the stability properties of the trivial steady state. Then, the linearized LBE becomes

$$f_i(\mathbf{r}+m\mathbf{c}_i, k+\tau) - f_i(\mathbf{r},k) = \sum_{j=1}^{\tilde{b}} \mathbf{\Omega}_{ij} f_j(\mathbf{r},k) + \frac{1}{\tilde{b}} \sum_{j=1}^{\tilde{b}} \left.\frac{\partial \bar{\mathcal{R}}_i}{\partial f_j}\right|_{\bar{f}=0}(\mathbf{r},k) f_j(\mathbf{r},k), \qquad (4.53)$$

where the linearized growth term reads

$$\left.\frac{\partial \bar{\mathcal{R}}_i}{\partial f_j}\right|_{\bar{f}=0}(\mathbf{r},k) = \frac{r_m}{\tilde{b}}. \qquad (4.54)$$

Now, since the system (4.53) is linear, we can introduce the discrete Fourier transform with wavenumber $\mathbf{q} = (q_1, q_2)$ of the corresponding Fourier mode:

$$f_i(\mathbf{r},k) = A^{\tau k} e^{i\langle \mathbf{q}, \mathbf{c}_i \rangle m} \hat{f}_i, \qquad (4.55)$$

where $\langle \cdot, \cdot \rangle$ is the inner product of two vectors. Then, from the system (4.52), we obtain the following algebraic set of equations for the $\mathbf{f}_i$'s:

$$\mathbf{M}\,\hat{\mathbf{f}} = 0, \qquad (4.56)$$

where the matrix $\mathbf{M}$ is the Fourier transform of the matrix $\mathbf{\Gamma}$, which is called Boltzmann propagator, and $\hat{\mathbf{f}} = (\hat{f}_1,...,\hat{f}_{\tilde{b}})$. The elements of $\mathbf{M}$ are:

4.3. Mean-field analysis

$$\mathbf{M}_{ij} = \frac{1}{\tilde{b}}(1+r_m) - A^\tau e^{i\langle \mathbf{q},\mathbf{c}_i\rangle m}\delta_{ij}, \quad (4.57)$$

where $i,j = 1, ..., \tilde{b}$. A non-trivial solution exists if $det(\mathbf{M}) = 0$. Making explicit use of this condition, we obtain a $\tilde{b}^{th}$ order polynomial equation for the damping coefficient A

$$\frac{2}{\tilde{b}}A^{\tau(\tilde{b}-1)}\left[(1+r_m)(\cos(mq_1)+\cos(mq_2)) + \frac{\tilde{b}}{2}A^\tau + (\frac{\tilde{b}}{2}-2)(1+r_m)\right] = 0. \quad (4.58)$$

The solutions of A for the above discrete dispersion relation are:

$$A^\tau_{(1)}(\mathbf{q}) = 2(1+r_m)\left[\frac{\cos(mq_1)+\cos(mq_2)}{\tilde{b}} + 1 - \frac{4}{\tilde{b}}\right], \quad (4.59)$$

$$A^\tau_{(j)}(\mathbf{q}) = 0, \text{ for } j = 2, ..., \tilde{b}. \quad (4.60)$$

The damping coefficient $A^\tau_{(1)}$ depends on $\mathbf{q}$ and its value is larger than one if $r_m > 0$. Moreover, the leading damping coefficient can be expressed as the exponential of an equivalent continuous damping rate $z(\mathbf{q})$, i.e. $A^\tau_{(1)}(\mathbf{q}) = e^{z(\mathbf{q})}$ or $z(\mathbf{q}) = \tau ln(A_{(1)}(\mathbf{q}))$. After this transformation of the damping coefficient $A^\tau_{(1)}$ we obtain

$$z(\mathbf{q}) = \frac{1}{\tau}ln(1+r_m) - \frac{m^2}{\tilde{b}\tau}|\mathbf{q}|^2 + \mathcal{O}(\mathbf{q}^4). \quad (4.61)$$

Using the continuous, inverse Fourier transform and assuming small growth rates $r_m \ll 1$, we can obtain a linear reaction-diffusion equation (RDE) that describes the macroscopic spatio-temporal evolution of the system around the homogeneous state $\bar{f} = 0$ corresponding to $\rho_0 = 0$:

$$\partial_t \rho = \frac{m^2}{\tilde{b}\tau}\nabla^2 \rho + \frac{r_m}{\tau}\rho. \quad (4.62)$$

The above macroscopic equation is a linearized version of eq. (4.36) around the steady state $\rho_0 = 0$, i.e.

$$\partial_t \rho = \frac{m^2}{\tilde{b}\tau}\nabla^2 \rho + \frac{1}{\tau}\frac{\partial F(\rho)}{\partial \rho}\bigg|_{\rho=0}\rho, \quad (4.63)$$

where $F(\rho)$ is defined in eq. (4.37).

4.3.2.4 Discussion of scaling methods

In order to gain insight into the spatio-temporal behavior of our LGCA growth process, we have derived the corresponding macroscopic description using three different scaling methods (the Chapman-Enskog, the CML and the Fourier space method). Since the main focus of the thesis is not the critical review of all scaling methods, we provide a short discussion of the above-presented methods. In particular, we focus on the discussion of the kind of macroscopic description, the consideration of non-linearities and the range of validity for particular parameters (r_m):

	Macroscopic eq.	Non-linearities	Range of validity
Chapman-Enskog	RDE	Yes	$r_m \ll 1$
CML method	Integro-difference eq.	Yes	$\forall r_m \in [0,1]$
Fourier method	RDE	No	$r_m \ll 1$

Table 4.1: Here, we summarize the characteristics of the different scaling methods employed to derive the macroscopic descriptions of our growth LGCA. For more details see in text.

Macroscopic equation All three methods are able to provide a macroscopic description of our growth LGCA. However, these macroscopic equations are not completely identical. In particular the Chapman-Enskog method and the Fourier space method provide two RDE's ((4.36) and (4.62) respectively) as macroscopic limits of the process. In contrast, the CML method provides an integro-difference macroscopic equation (4.42). The reason is that in the CML method we do approximate sums by integrals and not by differential operators. For a specific choice of the scaling argument (4.27), eq. (4.42) can be represented also by the RDE (4.36).

Non-linearities Not all of the three scaling methods are able to capture the non-linearities of the growth process. The CML method is able to capture fully the non-linearities of the problem, which is stressed in eq. (4.46). The Chapman-Enskog method captures the non-linearities of the growth operator as evidenced in the eq. (4.36). Since the Fourier space method demands the system to be linear, it produces a linearized version (4.62) of eq. (4.36).

Range of validity Finally, we discuss the range of parameters that ensures the validity of the approximation. The CML method produces a condition-free macroscopic description, valid for the full range of the LGCA's parameters. However, the validity of the other two methods is restricted by the condition of very low reproductive rates, i.e. $r_m \ll 1$. In the case of Chapman-Enskog method, reproduction rates are required to be small since the dominant process is diffusion and the cell doubling is considered act as a perturbation. In the case of Fourier space method, the reason is the linearization of the dispersal eq. (4.61) around $r_m = 0$.

Table 4.1 summarizes the characteristics of the above mentioned methods. In the following chapters, we use one of the above methods depending on the demands of the problem.

4.4 Mean-field approximations and macroscopic behavior

In the previous section, we have presented two kinds of mean-field approximations of our LGCA, i.e. the temporal and the spatiotemporal. Consequently, one would wonder which approach is more appropriate to describe the automaton's macroscopic behavior. In this section, we provide a comparison of the two mean-field approaches and we attempt to identify their limitations. In particular, we use the well-stirred and the spatially distributed mean-field approximations to gain insight into: (i) the spatio-temporal pattern formation of the LGCA and (ii) the behavior of important macroscopic observables, such as *total number of cells* ($\bar{\rho}(t)$) and the *per capita growth rate* (γ).

From the mean-field analysis of the well-stirred system, we have derived an ODE (4.23) that allows us to calculate the net change in the node density for any given time. The total number of cells in the system is $\bar{\rho}(t) = \sum_{\mathbf{r} \in \mathcal{L}} \rho(t) = |\mathcal{L}|\rho(t)$. Thus, using eq. (4.23), the time evolution of the total number of cells in the system is given by

$$\bar{\rho}(t) = \frac{C}{1 + \left(\frac{C}{\bar{\rho}(0)} - 1\right)e^{-\frac{r_m}{\varepsilon\tau}t}}, \qquad (4.64)$$

where $\bar{\rho}(0)$ is the initial total number of cells on the lattice and $C = \tilde{b}|\mathcal{L}|$ the total capacity of the lattice $\mathcal{L}$. The per capita growth rate is defined as the average number of offsprings per individual for a given time, i.e.

$$\gamma = \frac{\Delta\bar{\rho}(t)}{\bar{\rho}(t)}. \qquad (4.65)$$

Obviously, the total number of produced cells is $\Delta\bar{\rho}(t) = d\bar{\rho}(t)/dt = r_m\bar{\rho}(1 - \frac{\bar{\rho}}{\tilde{b}})$. Therefore, the per capita growth rate for the well-stirred system is given by:

$$\gamma_{tMF} = r_m(1 - \frac{\bar{\rho}}{\tilde{b}}). \qquad (4.66)$$

The well-stirred MF approximation is inappropriate for the analysis of spatio-temporal pattern formation, since space is not considered in eq. (4.23).

Now, we demonstrate how the macroscopic description derived from the spatially distributed case (4.37) allows for an analysis of the spatio-temporal pattern formation in the growth LGCA.

The evolving pattern observed in simulations, starting from an initial fully occupied cluster of nodes, is an isotropically growing disc (fig. 4.1). The isotropy of the system becomes apparent in the Chapman-Enskog expansion (4.38), in particular by comparing the mean fluxes of cells along x-y directions and calculating the total mean flux [Hatzikirou 2008b], i.e.

$$\left|\langle \mathbf{J}_{x+}\rangle - \langle \mathbf{J}_{y+}\rangle\right| = |\mathbf{c_1}f_1 - \mathbf{c_2}f_2| = 0 \wedge \langle \mathbf{J}\rangle = \sum_{i=1}^{\tilde{b}} \mathbf{c}_i f_i = 0, \qquad (4.67)$$

Chapter 4. Growth processes

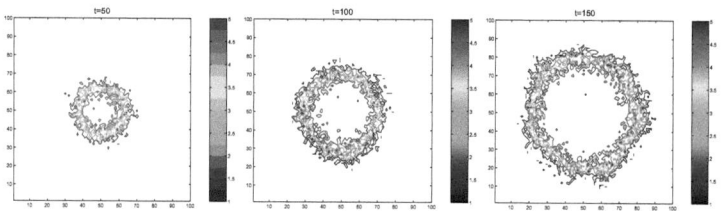

Figure 4.1: Typical simulations of the spatiotemporal evolution of the LGCA growth process starting from an initial fully occupied cluster of nodes in the center of the lattice. The three figures show snapshots of the same simulation at different times. The colors encode the node density.

where the *flux* is defined as

$$\mathbf{J}(\boldsymbol{\eta}(\mathbf{r},k)) = \sum_{i=1}^{b} \mathbf{c}_i \eta_i(\mathbf{r},k).$$

Additionally, simulations indicate a moving front along which the occupancy of the initially empty nodes is increasing from zero cells to the maximum capacity $\tilde{b}$. This behavior is predicted by the FK eq. (4.7), which is known that results in a front where the stable phase (fully occupied nodes) propagates towards the unstable one (empty nodes).

In order to predict the total number of cells and the per capita growth rate, we assume that (i) the system evolves for asymptotically long times and (ii) the initial front is sufficiently steep. Under these assumptions, the speed of the front relaxes to an asymptotic value $c^* = c_{min} = 2\sqrt{\frac{m^2}{\tilde{b}}r_m}$ [Murray 2001]. Additionally, the above assumption allows us to consider that the resulting front is extremely steep (sharp interface between the stable and the unstable phase).

Now, we can calculate the total number of cells from the relation, $\bar{\rho}(t) = \tilde{b}\pi R^2(t)$ (the mass of the disc), where the radius grows as $R(t) = c^*t$. Thus, the total number of cells becomes:

$$\bar{\rho}(t) = 4\pi m^2 r_m t^2. \qquad (4.68)$$

The per capita growth rate is given by the relation $\gamma = \Delta\bar{\rho}(t)/\bar{\rho}(t)$, where $\Delta\bar{\rho}(t) = d\bar{\rho}(t)/dt = 8\pi m^2 r_m t$ is the change of $\bar{\rho}(t)$ within one time step. Thus, it follows that

$$\gamma_{spMF} = \frac{4m\sqrt{\pi r_m}}{\sqrt{\bar{\rho}(t)}} \sim \bar{\rho}(t)^{-1/2}. \qquad (4.69)$$

In fig. 4.2, we show the behavior of the per capita growth of both mean-field approximations in comparison with actual LGCA simulation results. We observe that

4.5. Summary

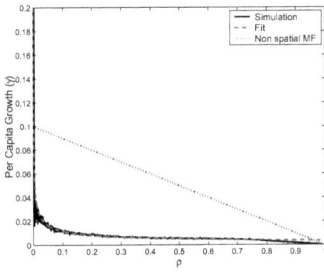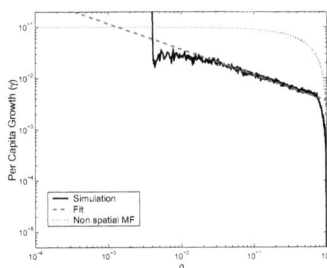

Figure 4.2: Evolution of the per capita growth γ as a function of the total population density $\bar{\rho}$. Observe that the γ calculated from the LGCA simulations, decreases rapidly for increasing population densities. The behavior of γ can be fitted by a curve $A\bar{\rho}^{-1/2}$, as it is calculated in our spatio-temporal MF analysis (4.69). Additionally, we observe that the temporal MF (4.66) completely fails to follow the actual LGCA dynamics. The log-log plot (right figure) allows for a better distinction between the fit and the simulation curves, especially for low $\bar{\rho}$.

γ_{tMF} fails completely to match the per capita growth of the simulations. The per capita growth γ_{spMF}, derived from the spatio-temporal mean-field, offers a better insight into the actual growth dynamics, since it recovers qualitatively the scaling (4.69) of the per capita growth rate compared to the one calculated from the simulations. However, the γ_{spMF} does not agree quantitatively to the actual per capita growth.

4.5 Summary

In this chapter, the focus is on the modeling and the mathematical analysis of growth processes using LGCA. In particular, we are interesting in the mathematical analysis of growth process macroscopic behavior, which can be achieved by means of mean-field approximation. The idea of the mean-field approximation is the reduction of the description of a system with many interacting individuals (large degree of freedom), like CA, to the level of an effective, average description of the behavior of a single individual (low degree of freedom). This reduction can be achieved in different ways. Here in particular, the mean-field approximation neglects all the spatial correlations for a given time step of a CA. This assumption facilitates the calculation of the node density distribution of individuals over a lattice. In particular, the mean-field approximation can be derived under the assumption of (i) a well-stirred system which leads to a spatially homogeneous node density distribution and (ii) a spatially distributed system which leads to an equation that allows us to calculate a spatially inhomogeneous node density distribution. Both kinds of mean-field approximations allow for an effective, macroscopic description of the behavior of a given CA. In particular, the well-stirred case leads to a temporal mean-field description, since the spatial resolution is lost. On the contrary, the spatially distributed case allows for the derivation of a spatio-temporal mean-field approximation of the actual CA dynamics. In order to gain insight into the spatio-temporal behavior of our LGCA growth process, we have derived the corresponding macroscopic descriptions. For the spatially distributed mean-field approximation case, we have used three scaling methods, the Chapman-Enskog, the CML and the Fourier space method. The characteristics of the above mentioned methods are summarized in table 4.1. In contrast to the well-stirred case, the spatially distributed case provides a satisfactory description of the system's spatial pattern formation and recovers the scaling laws of important macroscopic measures, like per capita growth rate.

Altogether, the derived spatio-temporal mean-field description is an insightful approximation of the stochastic growth process. Although this spatio-temporal mean-field approximation qualitatively and quantitatively agrees with most of the system's macroscopic behavior, it fails to provide exact quantitative predictions of those properties which rely on higher order spatial correlations. This is due to the nature of the mean-field approximation, since it neglects the spatial correlations, and the sharp interface assumption. The further development and extension of the mean-field approach opens a challenging research field but is beyond the scope of this thesis [Hatzikirou 2008a].

Part II

Tumor invasion

CHAPTER 5

The impact of environment on tumor cell migration

Contents

5.1	Introduction	**61**
	5.1.1 Types of cell motion	62
	5.1.2 Mathematical models of cell migration	64
	5.1.3 Overview of the chapter	65
5.2	LGCA models of cell motion in a static environment	**65**
	5.2.1 Model I	70
	5.2.2 Model II	70
5.3	Analysis of the LGCA models	**74**
	5.3.1 Model I	74
	5.3.2 Model II	78
5.4	Summary	**80**

5.1 Introduction

Alan Turing in his landmark paper of 1952 introduced the concept of self-organization to biology [Turing 1952]. He suggested "that a system of chemical substances, called morphogens, reacting together and diffusing through a tissue, is adequate to account for the main phenomena of morphogenesis". Such a system, although it may originally be quite homogeneous, may later develop a pattern or structure due to an instability of the homogeneous equilibrium, which is triggered off by random disturbances. Today, it is realized that, in addition to diffusible signals, the role of cells in morphogenesis is crucial. In particular, living cells possess migration strategies that go far beyond the merely random displacements characterizing non-living molecules (diffusion). It has been shown that the microenvironment plays an important role in the way that cells select their migration strategies [Friedl 2000]. Moreover, the microenvironment provides the prototypic substrate for cell migration in embryonic morphogenesis, immune defense, wound repair or *tumor invasion* (see Introduction).

62   Chapter 5. The impact of environment on tumor cell migration

The cellular microenvironment is a highly heterogeneous medium for cell motion including the extracellular matrix composed of fibrillar structures, collagen matrices, diffusible chemical signals as well as other mobile and immobile cells. Cells move within their environment by responding to various stimuli. In addition, cells change their environment locally by producing or absorbing chemicals and/or by degrading the neighboring tissue. This interplay establishes a dynamic relationship between individual cells and the surrounding substrate. In the following subsection, we provide more details about the different cell migration strategies in various environments. Environmental heterogeneity contributes to the complexity of the resulting cellular behaviors. Moreover, cell migration and interactions with the environment are taking place at different spatiotemporal scales. Mathematical modeling has proven extremely useful in getting insights into such multiscale systems.

In this chapter, we show how a suitable microscopical mathematical model (a LGCA) can contribute to understand the interplay of moving cells with their heterogeneous environment. Here, we are interest to reveal how tumor's environment influences the tumor cell motion. However, the discussed models of this chapter can be extended in healthy cell motion in different environments, since we assume that it is completely analogous to the tumor cell case. In particular, in order to understand the phenomenon of cell motion into different environments, we address the following questions:

- What kind of spatio-temporal patterns are formed by moving cells using different strategies?

- How does the moving cell population affect its environment and what is the feedback to its motion?

- What is the spreading speed of a cell population within a heterogeneous environment?

5.1.1   Types of cell motion

The cell migration type is strongly coupled to the kind of environment that hosts the cell population. A range of external cues impart information to the cells which regulate their movement, including long-range diffusible chemicals (e.g. chemoattractants), contact with membrane-bound molecules on neighboring cells (mediating cell-cell adhesion and contact inhibition) and contact with the extracellular matrix (ECM) surrounding the cells (contact guidance, haptotaxis). Accordingly, the environment can act on the cell motion in many different ways.

Recently, Friedl et al. [Friedl 2000, Friedl 2004] have investigated in depth different kinds of cell movement in tissues. The main processes that influence cell motion are identified by: cell-ECM adhesion forces introducing integrin-induced motion and cell-cell adhesive forces leading to cadherin-induced motion. The different contributions of these two kinds of adhesive forces characterize the particular type of cell motion. Table 5.1 gives an overview of the possible types of cell migration in the ECM.

5.1. Introduction

Type/Motion	Random walk	Cell-Cell Adhesion	Cell-ECM Adhesion	Proteolysis
Amoeboid	++	-/+	-/+	+/-
Mesenchymal	-	-/+	+	+
Collective	-	++	++	+

Table 5.1: Diversity in cell migration strategies (after Friedl et al. [Friedl 2004]). In different tissue environments, different cell types exhibit either individual (amoeboid or mesenchymal) or collective migration mechanisms to overcome and integrate into tissue scaffolds (see text for explanations).

Amoeboid motion is the simplest kind of cell motion and can be characterized as random motion of cells without being affected by the integrin concentration of the underlying matrix. Amoeboidly migrating cells develop a dynamic leading edge rich in small pseudopodia, a roundish or ellipsoid main cell body and a trailing small uropod (rear edge of a moving cell). Cells, like neutrophils, perceive the tissue as a porous medium, where their flexibility allows them to move through the tissue without significantly changing it. On the other hand, mesenchymal motion of cells (for instance glioma cells) leads to alignment with the fibres of the ECM, since the cells are responding to environmental cues of non-diffusible molecules bound to the matrix and follow the underlying structure. Mesenchymal cells retain an adhesive, tissue-dependent phenotype and develop a spindle-shaped elongation in the ECM. In addition, the proteolytic activity (metalloproteinases production) of such cells allows for the remodeling of the matrix and establishes a dynamical environment. The final category is collective motion of cells (e.g. endothelial cells) that respond to cadherins and create cell-cell bounds. Clusters of cells can move through the adjacent connective tissue. Leading cells provide the migratory traction and, via cell-cell junctions, pull the following group forward.

One can think about two distinct ways of cells responding to environmental stimuli: either the cells are following a certain direction and/or the environment imposes an orientational preference. An example of the directed case is the graded spatial distribution of adhesion ligands along the ECM which is thought to influence the direction of cell migration [McCarthy 1984], a phenomenon known as haptotaxis [Carter 1965]. Chemotaxis mediated by diffusible chemotactic signals provides a further example of directed cell motion in a dynamically changing environment. On the other hand, alignment is observed in fibrous environments where amoeboid and mesenchymal cells change their orientation according to the fibre structure. Mesenchymal cells use additionally proteolysis to facilitate their movement and remodel the neighboring tissue (dynamic environment). Table 5.2 summarizes the above statements.

It has been shown that the basic strategies of cell migration are retained in tumor cells [Friedl 2003]. However, it seems that tumor cells can adapt their strategy, i.e. the cancer cell's migration mechanisms can be reprogrammed, allowing it to acquire invasive properties via morphological and functional dedifferentiation [Friedl 2003]. Furthermore, it has been demonstrated that the microenvironment is crucial for

64   Chapter 5. The impact of environment on tumor cell migration

	Static	Dynamic
Direction	Haptotaxis	Chemotaxis
Orientation	Amoeboid	Mesenchymal

Table 5.2: In this table, we classify the environmental effect with respect to different cell migration strategies. One can distinguish static and dynamic environments. In addition, we identify environments that impart directional or only orientational information for migrating cells (see text for explanations).

cancer cell migration, e.g. fiber tracks in the brain's white matter facilitate glioma cell motion [Swanson 2002, Hatzikirou 2005]. Therefore, a better understanding of cell migration strategies in heterogeneous environments is particularly crucial for designing new cancer therapies.

5.1.2 Mathematical models of cell migration

This chapter focuses on the mathematical analysis of cell motion in heterogeneous environments. A large number of mathematical models has already been proposed to model various aspects of cell motion. Reaction-diffusion equations have been used to model the phenomenology of motion in various environments, like diffusible chemicals (Keller-Segel chemotaxis model [Keller 1971] etc.) and mechanical ECM stresses [Murray 1983]. Integro-differential equations have been introduced to model fibre alignment in [Dallon 2001]. Navier-Stokes equations and the theory of fluid dynamics provided insight into the "flow" of cells within complex environments, for instance in porous media [Byrne 2003]. However, the previous deterministic, continuous models describe cell motion at a macroscopic level neglecting the microscopical cell-cell and cell-environment interactions. Kinetic equations [Dolak 2005, Chauviere 2007] and especially transport equations [Othmer 1988, Dickinson 1993, Dickinson 1995, Hillen 2006] have been proposed as models of cell motion along tissues, at a mesoscopic level of description (the equations describe the behavior of cells within a small partition of space). Microscopical experimental data and the need to analyze populations consisting of a low number of cells call for models that describe the phenomena at the level of cell-cell interactions, e.g. interacting particle systems [Liggett 1985], cellular automata (CA) [Deutsch 2005], off-lattice Langevin methods [Grima 2007, Newman 2004], active Brownian particles [Schweitzer 2003, Peruani 2007] and other microscopic stochastic models [Othmer 1997, Okubo 2002].

In the following, we show that LGCA can serve as models for migrating cell populations. Additionally, their discrete nature allows for the description of the cell-cell and cell-environment interactions at the microscopic level of single cells but at the same time enables us to observe the macroscopical evolution of the whole population.

5.2. LGCA models of cell motion in a static environment

5.1.3 Overview of the chapter

In this chapter, we will explore the role of the environment (both in a directional and orientational sense) for cell movement. Moreover, we consider cells that lack metalloproteinase production (proteolytic proteins) and do not change the ECM structure, introducing static environments (additionally, we do not consider diffusible environments). Moreover, we will consider populations with a constant number of cells (no proliferation/death of cells) in time.

In section 5.2, we define LGCA models of moving cells in different environments that impart directional and orientational information for the moving cells. Furthermore, we provide a tensor characterization for the environmental impact to migrating cell populations. As an example, we present simulations of a proliferative cell population in a tensor field defined by clinical DTI data. This system can serve as a model of *in vivo* glioma cell invasion. In section 5.3, we show how mathematical analysis of the LGCA model can yield an estimate for the cell dispersion speed within a given environment. Finally, in section 5.4, we sum up the results and we discuss potential venues for analysis, extensions and applications.

5.2 LGCA models of cell motion in a static environment

In this section, we define two LGCA models that describe cell motion in static environments. We especially address two problems:

- How can we model the environment?

- How should the automaton rules be chosen to model cell motion in a specific environment?

As we already stated, we want to mathematically model biologically relevant environments. According to Table 5.2, we distinguish a "directional" and an "orientational" environment, respectively. The mathematical entity that allows for the modeling of such environments is called a *tensor field*. A tensor field is a collection of different tensors which are distributed over a spatial domain. A *tensor* is (in an informal sense) a generalized linear 'quantity' or 'geometrical entity' that can be expressed as a multi-dimensional array relative to a choice of basis of the particular space on which the tensor is defined. The intuition underlying the tensor concept is inherently geometrical: a tensor is independent of any chosen frame of reference.

The rank of a particular tensor is the number of array indices required to describe it. For example, mass, temperature, and other scalar quantities are tensors of rank 0; while force, momentum, velocity and further vector-like quantities are tensors of rank 1. A linear transformation such as an anisotropic relationship between velocity vectors in different directions (diffusion tensors) is a tensor of rank 2. Thus, we can represent an environment with directional information as a vector (tensor of rank 1) field. The geometric intuition for a vector field corresponds to an 'arrow' attached to each point of a region, with variable length and direction (fig. 5.1). The

66 Chapter 5. The impact of environment on tumor cell migration

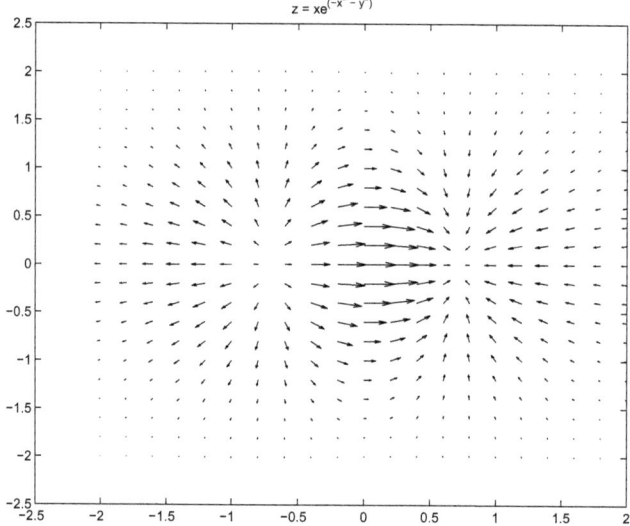

Figure 5.1: An example of a vector field (tensor field of rank 1). The vectors (e.g. integrin receptor density gradients) show the direction and the strength of the environmental drive.

5.2. LGCA models of cell motion in a static environment

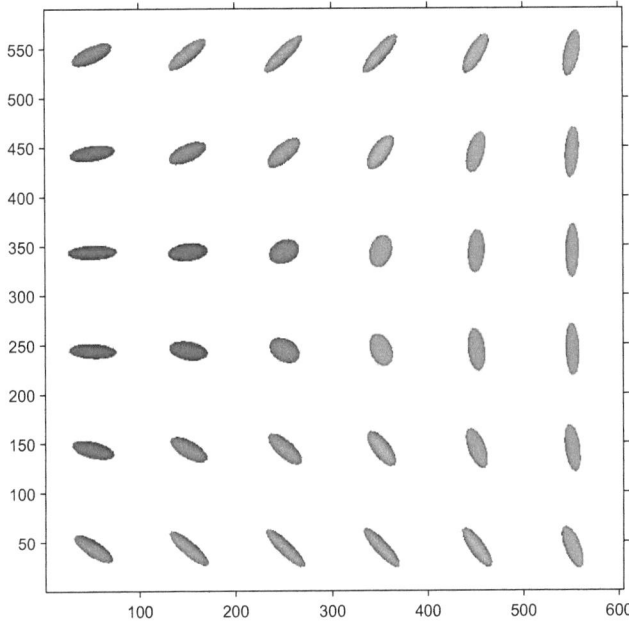

Figure 5.2: An example of a tensor field (tensor field of rank 2). We represent the local information of the tensor as ellipsoids. The ellipsoids can encode e.g. the degree of alignment of a fibrillar tissue. The colors are denoting the orientation of the ellipsoids.

idea of a vector field on a curved space is illustrated by the example of a weather map showing wind velocity, at each point of the earth's surface. An environment that carries orientational information for each geometrical point can be modeled by a tensor field of rank 2. A geometrical visualization of a second order tensor field can be represented as a collection of ellipsoids, assigned to each geometrical point (fig. 5.2). The ellipsoids represent the orientational information that is encoded into tensors.

To model cell motion in a given tensor field (environment)-of rank 1 or 2- we should modify the interaction rule of the LGCA. We use a special kind of interaction rules for the LGCA dynamics, firstly introduced by Alexander et al. [Alexander 1992]. We consider biological cells as random walkers that are reoriented by maximizing a potential-like term. Assuming that the cell motion is affected by cell-cell and cell-environment interactions, we can define the potential as the sum of

Chapter 5. The impact of environment on tumor cell migration

these two interactions.:

$$G(\mathbf{r}, k) = \sum_j G_j(\mathbf{r}, k) = G_{cc}(\mathbf{r}, k) + G_{ce}(\mathbf{r}, k), \quad (5.1)$$

where $G_j(\mathbf{r}, k)$, $j = cc, ce$ is the sub-potential that is related to cell-cell and cell environment interactions, respectively.

Interaction rules are formulated in such a way that cells preferably reorient into directions which maximize (or minimize) the potential, that is according to the gradients of the potential $\mathbf{G}'(\mathbf{r}, k) = \nabla G(\mathbf{r}, k)$.

Consider a lattice-gas cellular automaton defined on a two-dimensional lattice with b velocity channels ($b = 4$ or $b = 6$). Let the number of particles at node r at time k be denoted by

$$n(r, k) = \sum_{i=1}^{b} \eta_i(\mathbf{r}, k),$$

and the *flux* be denoted by

$$\mathbf{J}(\boldsymbol{\eta}(\mathbf{r}, k)) = \sum_{i=1}^{b} \mathbf{c}_i \eta_i(\mathbf{r}, k).$$

The probability that $\boldsymbol{\eta}^C$ is the outcome of an interaction at node r is defined by

$$\mathbb{P}(\boldsymbol{\eta} \to \boldsymbol{\eta}^C | G(\mathbf{r}, k)) = \frac{1}{Z} \exp\left[\alpha F\big(\mathbf{G}'(\mathbf{r}, k), \mathbf{J}(\boldsymbol{\eta}^C)\big)\right] \delta\big(n(\mathbf{r}, k), n^C(\mathbf{r}, k)\big), \quad (5.2)$$

where $\boldsymbol{\eta}$ is the pre-interaction state at $\mathbf{r}$ and the Kronecker's δ assumes the mass conservation of this operator. The sensitivity is tuned by the positive, real parameter α. The normalization factor is given by

$$Z = Z(\boldsymbol{\eta}(\mathbf{r}, k)) = \sum_{\boldsymbol{\eta}^C \in \mathcal{E}} \exp\left[\alpha F\big(\mathbf{G}'(\mathbf{r}, k), \mathbf{J}(\boldsymbol{\eta}^C)\big)\right] \delta\big(n(\mathbf{r}, k), n^C(\mathbf{r}, k)\big).$$

$F(\cdot)$ is a function that defines the effect of the $\mathbf{G}'$ gradients on the new configuration. A common choice of $F(\cdot)$ is the inner product $<,>$, which favors (or penalizes) the configurations that tend to have the same (or inverse) direction of the gradient $\mathbf{G}'$. Accordingly, the dynamics is fully specified by the following microdynamical equation (for more details see Chapter 2)

$$\eta_i(\mathbf{r} + \mathbf{c}_i, k + 1) = \eta_i^C(\mathbf{r}, k).$$

In the following, we present two stochastic potential-based interaction rules that correspond to the motion of cells in a vector field and a rank 2 tensor field, respectively. We exclude any other cell-cell interactions and we consider that the population has a fixed number of cells (mass conservation).

5.2. LGCA models of cell motion in a static environment 69

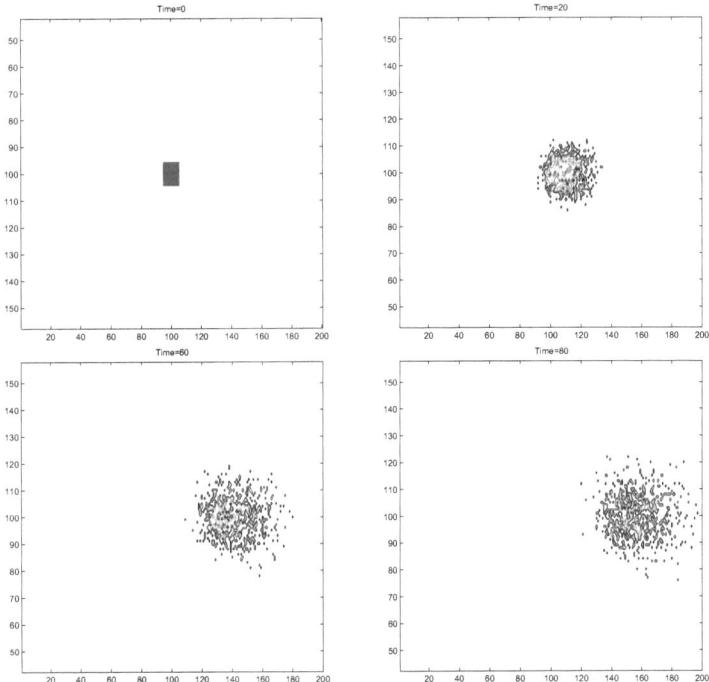

Figure 5.3: Time evolution of a cell population under the effect of a field $\mathbf{E}=(1,0)$. One can observe that the environmental drive moves all the cells of the cluster into the direction of the vector field. The blue color stands for low, the yellow for intermediate and red for high densities.

5.2.1 Model I

The first rule describes cell motion in a static environment that carries directional information expressed by a vector field **E**. Biologically relevant examples are the motion of cells that respond to fixed integrin concentrations along the ECM (haptotaxis). The spatial concentration differences of integrin proteins constitute a gradient field that creates a kind of "drift" **E** [Dickinson 1993]. We choose a two dimensional LGCA without rest channels and the stochastic interaction rule of the automaton follows the definition of the potential-based rules (eq. (5.1) with $\alpha = 1$):

$$\mathbb{P}(\eta \to \eta^C)(\mathbf{r},k) = \frac{1}{Z}\exp\left(\langle \mathbf{E}(\mathbf{r}), \mathbf{J}(\eta^C(\mathbf{r},k))\rangle\right)\delta\big(n(\mathbf{r},k), n^C(\mathbf{r},k)\big). \quad (5.3)$$

We simulate our LGCA for spatially homogeneous **E** for various intensities and directions. In fig. 5.3, we observe the time evolution of a cell cluster under the influence of a given field. We see that the cells collectively move towards the gradient direction and they roughly keep the shape of the initial cluster. The simulations in fig. 5.4 show the evolution of the system for different fields. It is evident that the "cells" follow the direction of the field and their speed responds positively to an increase of the field intensity.

5.2.2 Model II

We now focus on cell migration in environments that promote alignment (orientational changes). Examples of such motion are provided by neutrophil or leukocyte movement through the pores of the ECM, the motion of cells along fibrillar tissues or the motion of glioma cells along fiber track structures. As stated before, such an environment can be modeled by the use of a second rank tensor field that introduces a spatial anisotropy along the tissue. In each point, a tensor (i.e. a matrix) informs the cells about the local orientation and strength of the anisotropy and proposes a principle (local) axis of movement. For instance, the brain's fibre tracks impose a spatial anisotropy and their degree of alignment affects the strength of anisotropy.

Here, we use the information of the principal eigenvector of the diffusion tensor which defines the local principle axis of cell movement. Thus, we end up again with a vector field but in this case we exploit only the orientational information of the vector. The new rule for cell movement in an "oriented environment" is:

$$\mathbb{P}(\eta \to \eta^C)(\mathbf{r},k) = \frac{1}{Z}\exp\left(|\langle \mathbf{E}(\mathbf{r}), \mathbf{J}(\eta^C(\mathbf{r},k))\rangle|\right)\delta\big(n(\mathbf{r},k), n^C(\mathbf{r},k)\big). \quad (5.4)$$

In fig. 5.5, we show the time evolution of a simulation of model II for a given field. Fig. 5.6 shows the typical resulting patterns for different choices of tensor fields. We observe that the anisotropy leads to the creation of an ellipsoidal pattern, where the length of the main ellipsoid's axis correlates positively with the anisotropy strength.

This rule can be used to model the migration of glioma cells within the brain. Glioma cells tend to spread faster along fiber tracks. Diffusion Tensor Imaging

5.2. LGCA models of cell motion in a static environment

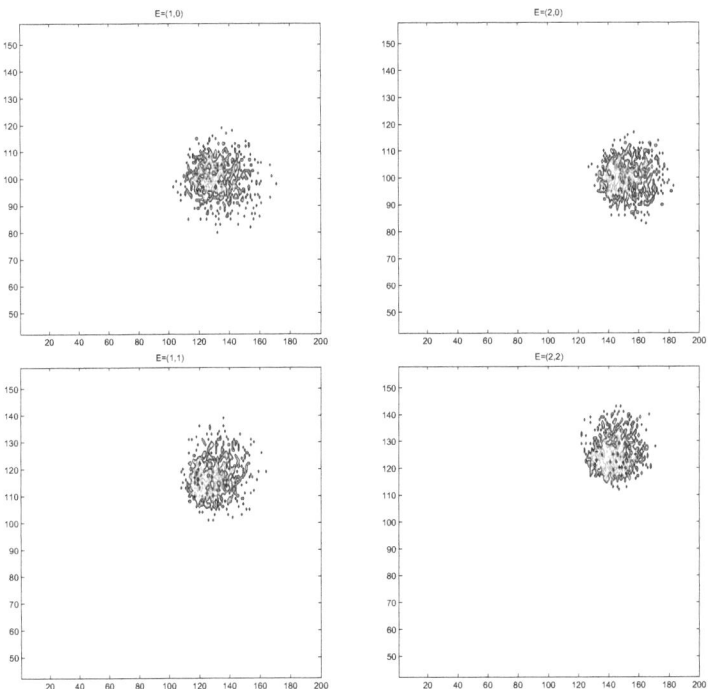

Figure 5.4: The figure shows the evolution of the cell population under the influence of different fields (100 time steps). Increasing the strength of the field, we observe that the cell cluster is moving faster in the direction of the field. This behavior is characteristic of a haptotactically moving cell population. The initial condition is a small cluster of cells in the center of the lattice. Colors denote different node densities (as in fig. 5.3).

72   Chapter 5. The impact of environment on tumor cell migration

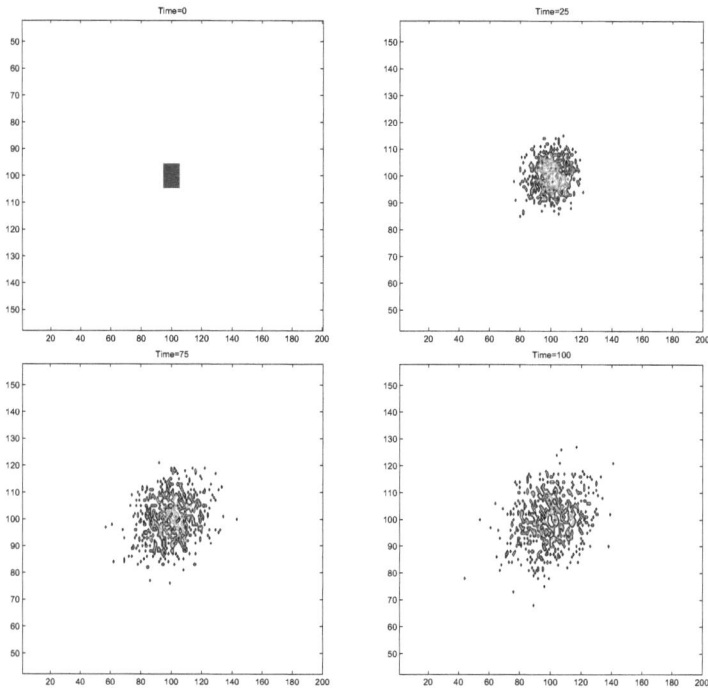

Figure 5.5: Time evolution of a cell population under the effect of a tensor field with principal eigenvector (principal orientation axis) **E**=(2,2). We observe cell alignment along the orientation of the axis defined by E, as time evolves. Moreover, the initial rectangular shape of the cell cluster is transformed into an ellipsoidal pattern with principal axis along the field **E**. Colors denote the node density (as in fig. 5.3).

5.2. LGCA models of cell motion in a static environment

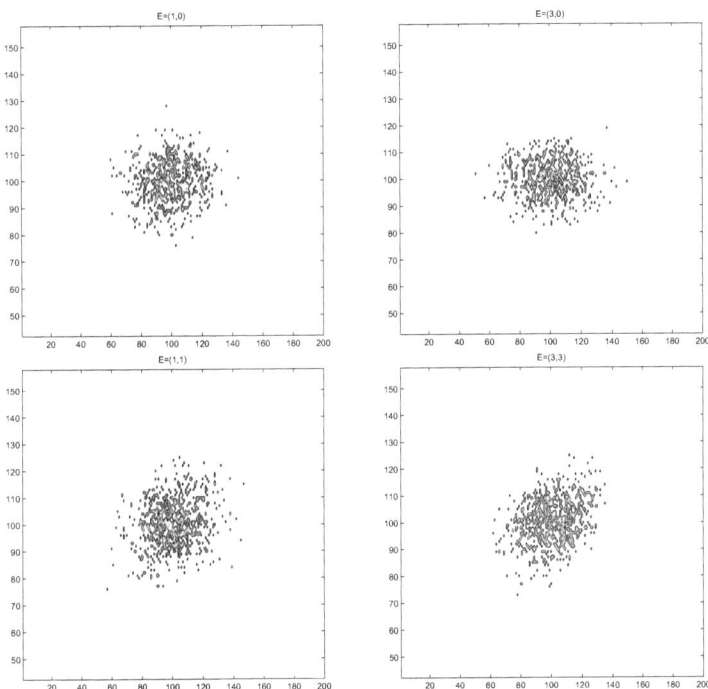

Figure 5.6: In this graph, we show the evolution of the pattern for four different tensor fields (100 time steps). We observe the elongation of the ellipsoidal cell cluster when the strength is increased. Above each figure the principal eigenvector of the tensor field is denoted. The initial conditions is always a small cluster of cells in the center of the lattice. The colors denote the density per node (as in fig. 5.3).

74   Chapter 5. The impact of environment on tumor cell migration

(DTI) is a Magnetic Resonance Imaging (MRI) based method that provides the local anisotropy information in terms of diffusion tensors. High anisotropy points belong to the brain's white matter, which consists of fiber tracks. A preprocessing of the diffusion tensor field can lead to the principle eigenvectors' extraction of the diffusion tensors, that provides us with the local principle axis of motion. By considering a proliferative cell population, as in [Hatzikirou 2008a], and using the resulting eigenvector field we can model and simulate glioma cell invasion. In fig. 5.7, we simulate an example of glioma growth and show the effect of fiber tracks in tumor growth using the DTI information.

5.3   Analysis of the LGCA models

In this section, we provide a theoretical analysis of the proposed LGCA models. Our aim is to calculate the equilibrium cell distribution and to estimate the speed of cell dispersion under environmental variations. Finally, we compare our theoretical results with the simulations.

5.3.1   Model I

In this subsection, we analyze model I and we derive an estimate of the cell spreading speed in dependence of the environmental field strength. The first idea is to choose a macroscopically accessible observable that can be measured experimentally. A reasonable choice is the mean lattice flux $\langle \mathbf{J}(\eta^C) \rangle_\mathbf{E}$, which characterizes the mean motion of the cells, with respect to changes of the field's strength $|\mathbf{E}|$:

$$\langle \mathbf{J}(\eta^C) \rangle_\mathbf{E} = \sum_i \mathbf{c}_i f_i^{eq}, \qquad (5.5)$$

where f_i^{eq}, $i = 1, ..., b$ is the equilibrium density distribution of each channel. Mathematically, this is the mean flux *response* to changes of the external vector field $\mathbf{E}$. The quantity that measures the linear response of the system to the environmental stimuli is called *susceptibility*:

$$\chi = \frac{\partial \langle \mathbf{J} \rangle_\mathbf{E}}{\partial \mathbf{E}}. \qquad (5.6)$$

It appears if we expand the mean flux in terms of small fields as:

$$\langle \mathbf{J} \rangle_\mathbf{E} = \langle \mathbf{J} \rangle_{\mathbf{E}=0} + \frac{\partial \langle \mathbf{J} \rangle_\mathbf{E}}{\partial \mathbf{E}} \mathbf{E} + O(\mathbf{E}^2). \qquad (5.7)$$

For the zero-field case, the mean flux is zero since the cells are moving randomly within the medium (diffusion). Accordingly, for small fields $\mathbf{E} = \begin{pmatrix} e_1 \\ e_2 \end{pmatrix}$ the linear approximation reads

$$\langle \mathbf{J} \rangle_\mathbf{E} = \frac{\partial \langle \mathbf{J} \rangle_\mathbf{E}}{\partial \mathbf{E}} \mathbf{E}.$$

5.3. Analysis of the LGCA models

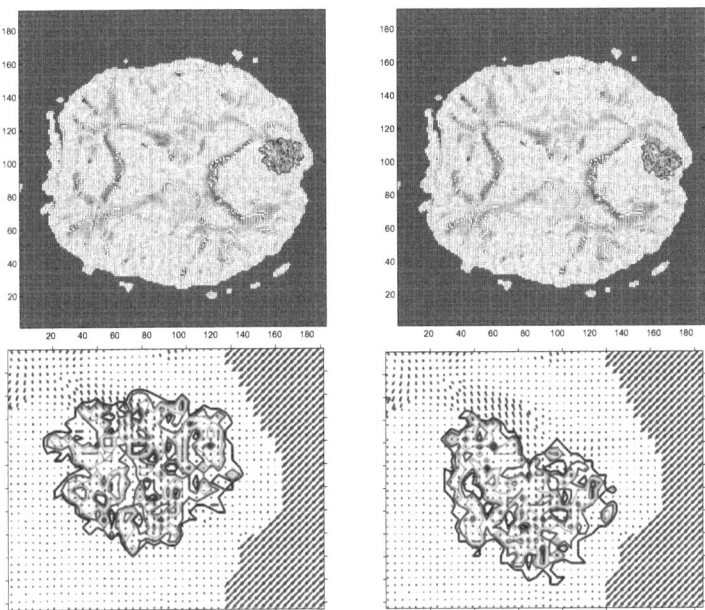

Figure 5.7: We show the brain's fiber track effect on glioma growth. We use a LGCA of a proliferating cancer cell population (for definition see [Hatzikirou 2008a]) moving in a tensor field provided by clinical DTI data, representing the brain's fiber tracks. **Top**: the left figure is a simulation without any environmental information (only diffusion). In the top right figure the effect of the fiber tracks in the brain on the evolution of the glioma growth is obvious. **Bottom**: The two figures display magnifications of the tumor area in the simulations above. This is an example of how environmental heterogeneity affects cell migration (where in this case tumor cell migration).

76　Chapter 5. The impact of environment on tumor cell migration

The *general linear response relation* is

$$\langle \mathbf{J}(\eta^C) \rangle_{\mathbf{E}} = \chi_{\alpha\beta} e_\beta = \chi e_\alpha, \tag{5.8}$$

where the second rank tensor is assumed to be isotropic, i.e. $\chi_{\alpha\beta} = \chi \delta_{\alpha\beta}$. In biological terms, we want to study the response of cell motion with respect to changes of the spatial distribution of the integrin concentration along the ECM, corresponding to changes in the resulting gradient field.

The aim is to estimate the stationary mean flux for fields $\mathbf{E}$. At first, we have to calculate the equilibrium distribution that depends on the external field. The external drive destroys the detailed balance (DB) conditions [1] that would lead to a Gibbs equilibrium distribution. In the case of non-zero external field, the system is out of equilibrium. The external field (environment) induces a breakdown of the spatial symmetry which leads to non-trivial equilibrium distributions depending on the details of the transition probabilities. The (Fermi) exclusion principle allows us to assume that the equilibrium distribution follows a kind of Fermi-Dirac distribution [Frisch 1987]:

$$f_i^{eq} = \frac{1}{1 + e^{x(\mathbf{E})}}, \tag{5.9}$$

where $x(\mathbf{E})$ is a quantity that depends on the field $\mathbf{E}$ and the mass of the system (if the DB conditions were fulfilled, the argument of the exponential would depend only on the invariants of the system). Moreover, the sigmoidal form of eq. (5.9) ensures the positivity of the probabilities $f_i^{eq} \geq 0$, $\forall x(\mathbf{E}) \in \mathbb{R}$. Thus, one can write the following *ansatz*:

$$x(\mathbf{E}) = h_0 + h_1 \mathbf{c}_i \mathbf{E} + h_2 \mathbf{E}^2. \tag{5.10}$$

After some algebra (the details can be found in Appendix 5.A), for small fields $\mathbf{E}$, one finds that the equilibrium distribution looks like:

$$f_i^{eq} = d + d(d-1)h_1 \mathbf{c}_i \mathbf{E} + \frac{1}{2}d(d-1)(2d-1)h_1^2 \sum_\alpha c_{i\alpha}^2 e_\alpha^2 + d(d-1)h_2 \mathbf{E}^2, \tag{5.11}$$

where $d = \rho/b$ and $\rho = \sum_{i=1}^b f_i^{eq}$ is the mean node density (which coincides with the macroscopic cell density) and the parameters h_1, h_2 have to be determined. Using the mass conservation condition, we find a relation between the two parameters (see Appendix 5.A):

$$h_2 = \frac{1 - 2d}{4} h_1^2. \tag{5.12}$$

---

[1] The detailed balance (DB) and the semi-detailed balance (SDB) imposes the following condition for the microscopic transition probabilities: $\mathbb{P}(\eta \to \eta^C) = \mathbb{P}(\eta^C \to \eta)$ and $\forall \eta^C \in \mathcal{E} : \sum_\eta \mathbb{P}(\eta \to \eta^C) = 1$. Intuitively, the DB condition means that the system jumps to a new micro-configuration and comes back to the old one with the same probability (micro-reversibility). The relaxed SDB does not imply this symmetry. However, SDB guarantes the existence of steady states and the sole dependence of the Gibbs steady state distribution on the invariants of the system (conserved quantities).

5.3. Analysis of the LGCA models

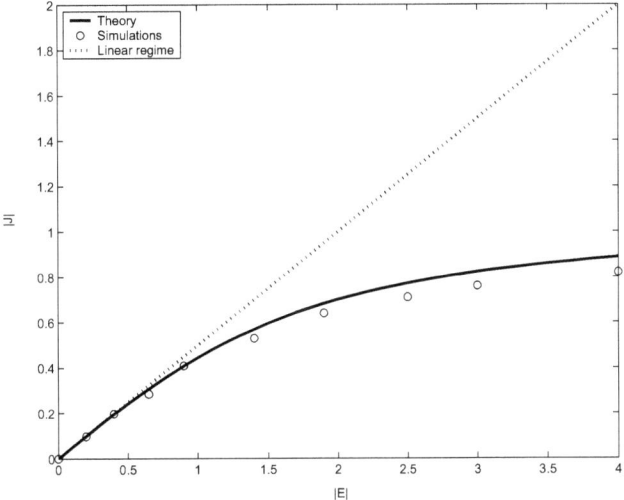

Figure 5.8: This figure shows the variation of the normalized measure of the total lattice flux $|\mathbf{J}|$ against the field intensity $|\mathbf{E}|$, where $\mathbf{E} = (e_1, e_2)$. We compare the simulated values with the theoretical calculations (for the linear and non-linear theory). We observe that the linear theory predicts the flux strength for low field intensities. Using the full distribution, the theoretical flux is close to the simulated values also for larger field strengths.

Finally, the equilibrium distribution can be explicitly calculated for small driving fields:
$$f_i^{eq} = d + d(d-1)h_1 \mathbf{c}_i \mathbf{E} + \frac{1}{2}d(d-1)(2d-1)h_1^2 Q_{\alpha\beta} e_\alpha e_\beta, \tag{5.13}$$
where $Q_{\alpha\beta} = c_{i\alpha} c_{i\beta} - \frac{1}{2}\delta_{\alpha\beta}$ is a second order tensor.

If we calculate the mean flux, using the equilibrium distribution up to first order terms of $\mathbf{E}$, we obtain from eq. (5.5) the linear response relation:
$$\langle \mathbf{J}(\eta^C) \rangle = \sum_i c_{i\alpha} f_i^{eq} = \frac{b}{2}d(d-1)h_1 \mathbf{E}. \tag{5.14}$$

Thus, the susceptibility reads:
$$\chi = \frac{1}{2}bd(d-1)h_1 = -\frac{1}{2}bg_{eq}h_1, \tag{5.15}$$
where $g_{eq} = f_i^{eq}(1 - f_i^{eq})$ is the equilibrium single particle fluctuation. In [Bussemaker 1996], the equilibrium distribution is directly calculated from the non-linear lattice Boltzmann equation corresponding to a LGCA with the same rule for

78 Chapter 5. The impact of environment on tumor cell migration

small external fields. In the same work, the corresponding susceptibility is determined and this result coincides with ours for $h_1 = -1$. Accordingly, we consider that $h_1 = -1$ in the following.

Our method allows us to proceed beyond the linear case, since we have explicitly calculated the equilibrium distribution of our LGCA:

$$f_i^{eq} = \frac{1}{1 + \exp\left(ln(\frac{1-d}{d}) - \mathbf{c}_i \mathbf{E} + \frac{1-2d}{4}\mathbf{E}^2\right)}. \tag{5.16}$$

Using the definition of the mean lattice flux eq. (5.5), we can obtain a good theoretical estimation for larger values of the field. Fig. 5.8 shows the behavior of the system's normalized flux obtained by simulations and a comparison with our theoretical findings. For small values of the field intensity $|\mathbf{E}|$ the linear approximation performs rather well and for larger values the agreement of our non-linear estimate with the simulated values is more than satisfactory. One observes that the flux response to large fields saturates. This is a biologically justified result, since the speed of cells is finite and an infinite increase of the field intensity should not lead to infinite fluxes (the mean flux is proportional to the mean velocity). Experimental findings in systems of cell migration mediated by adhesion receptors, such as ECM integrins, support the model's behavior [Palecek 1997, Zama 2006].

5.3.2 Model II

In the following section, our analysis characterizes cell motion by a different measurable macroscopic variable and provides an estimate of the cell dispersion for model II. In this case, it is obvious that the average flux, defined in (5.5), is zero (due to the symmetry of the interaction rule). In order to measure the anisotropy, we introduce the flux difference between $\mathbf{v}_1$ and $\mathbf{v}_2$, where the $\mathbf{v}_i$'s are eigenvectors of the anisotropy matrix (they are linear combinations of $\mathbf{c}_i$'s). For simplicity of the calculations, we consider $b = 4$ and X-Y anisotropy. We define:

$$|\langle \mathbf{J}_{\mathbf{v}_1} \rangle - \langle \mathbf{J}_{\mathbf{v}_2} \rangle| = |\langle \mathbf{J}_{x^+} \rangle - \langle \mathbf{J}_{y^+} \rangle| = |c_{11} f_1^{eq} - c_{22} f_2^{eq}|. \tag{5.17}$$

As before, we expand the equilibrium distribution around the field $\mathbf{E} = \mathbf{0}$ and we obtain equation

$$f_i = f_i(\mathbf{E} = \mathbf{0}) + (\nabla_\mathbf{E}) f_i \mathbf{E} + \frac{1}{2}\mathbf{E}^T(\nabla_\mathbf{E}^2) f_i \mathbf{E}. \tag{5.18}$$

With similar arguments as for the previous model I, we can assume that the equilibrium distribution follows a kind of Fermi-Dirac distribution (compare with eq. (5.9)). This time our *ansatz* has the following form,

$$x(\mathbf{E}) = h_0 + h_1|\mathbf{c}_i\mathbf{E}| + h_2\mathbf{E}^2, \tag{5.19}$$

because the rule is symmetric under the inversion $\mathbf{c}_i \to -\mathbf{c}_i$. Conducting similar calculations (Appendix 5.B) as in the previous subsection, one can derive the following expression for the equilibrium distribution:

5.3. Analysis of the LGCA models

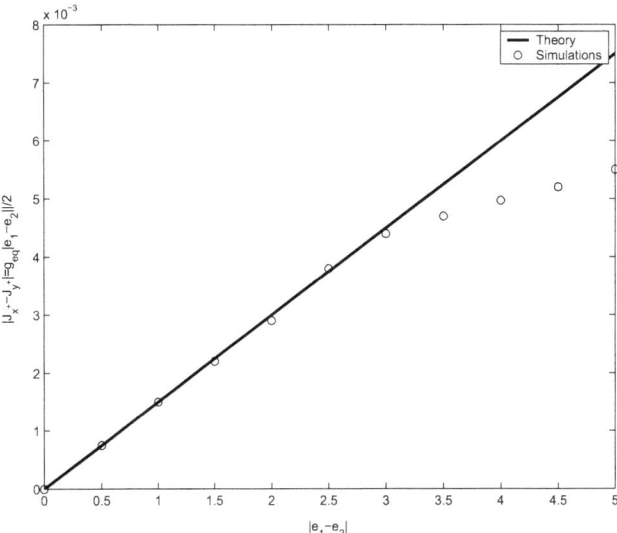

Figure 5.9: The figure shows the variation of the X-Y flux difference against the anisotropy strength (according to Model II). We compare the simulated values with the linear theory and observe a good agreement for low field strength. The range of agreement, in the linear theory, is larger than in the case of model I.

$$\begin{aligned} f_i^{eq} &= d + d(d-1)h_1|\mathbf{c}_i\mathbf{E}| \\ &+ \frac{1}{2}d(d-1)(2d-1)h_1^2 \sum_\alpha c_{i\alpha}^2 e_\alpha^2 \\ &+ d(d-1)(2d-1)h_1^2|c_{i\alpha}c_{i\beta}|e_\alpha e_\beta \\ &+ d(d-1)h_2\mathbf{E}^2. \end{aligned} \quad (5.20)$$

In Appendix 5.B, we identify a relation between h_1 and h_2 using the microscopic mass conservation law. To simplify the calculations we assume a square lattice (similar calculations can also be carried out for the hexagonal lattice case) and using $c_{11} = c_{22} = 1$, we derive the difference of fluxes along the X-Y axes (we restrict ourselves here to the linear approximation):

$$|f_1^{eq} - f_2^{eq}| = d(d-1)h_1\left|\sum_\alpha |c_{1\alpha}|e_\alpha - \sum_\alpha |c_{2\alpha}|e_\alpha\right| = d(d-1)h_1|e_1 - e_2|. \quad (5.21)$$

80   Chapter 5. The impact of environment on tumor cell migration

We observe that the parameter h_1 is still free and we should find a way to calculate it. In Appendix 5.C, we use a method similar to the work of [Bussemaker 1996] and we find that $h_1 = -1/2$. Substituting this value into the last relation and comparing with simulations (fig. 5.9), we observe again a very good agreement between the linear approximation and the simulations.

5.4 Summary

In this study, our first goal was to interpret in mathematical terms the environment related to cell migration. In particular, we are interested in the influence of environment on tumor cell migration, which is considered analogous to healthy cell motion. We have distinguished both static and dynamic environments, depending on the interactions with the cell populations. Mathematical entities called tensors enable us to extract local information about the local geometrical structure of the tissue. Technological advances, like DTI (Diffusion Tensor Imaging), in image analysis allow us to identify the microstructure of *in vivo* tissues. The knowledge of the microenvironment gives us a detailed picture of the medium through which the cells move, at the cellular length scale. Microscopical models are able to exploit this micro-scaled information and capture the dynamics.

To study and analyze the effects of the microenvironment on cell migration, we have introduced a microscopical modeling method called LGCA. We have identified and modeled the two main effects of static environments on cell migration:

- The first model addresses motion in an environment providing directional information. Such environments can be mediated by integrin density gradient fields or diffusible chemical signals leading to haptotactical or chemotactical movement, respectively. We have carried out simulations for different static fields, in order to understand the environmental effect on pattern formation. The main conclusion is that such an environment favors the collective motion of the cells in the direction of the gradients. Interestingly, we observe in fig. 5.3 that the cell population coarsely keeps the shape of the initial cluster and moves towards the same direction. This suggests that collective motion is not necessary an alternative cell migration strategy, as described in [Friedl 2004]. According to our results, collective motion can be interpreted as emergent behavior in a population of amoeboidly moving cells in a directed environment. Finally, we have calculated theoretically an estimator of the cell spreading speed, i.e. the mean flux for variations of the gradient field strength. The results exhibit a positive response of the cell flux to increasing field strength. The saturation of the response for large stimulus emphasizes the biological relevance of the model.

- The second model describes cell migration in an environment that influences the orientation of the cells (e.g. alignment). Fibrillar ECMs induce cell alignment and can be considered as an example of an environment that affects cell

5.4. Summary

orientation. Simulations show that such motion produces alignment along a principal orientation (i.e. fiber) and the cells tend to disperse along it (fig. 5.5). Like model I, we have calculated the cell response to variations of the field strength, in terms of the flux difference between the principal axis of motion and its perpendicular. This difference gives us an estimate of the speed and the direction of cell dispersion. Finally, we observe a similar saturation plateau for large fields, as in model I. Moreover, we gave an application of the second model for the case of brain tumor growth using DTI data (fig. 5.7).

As we have shown, the environment can influence cell motion in different ways. An interesting observation is that directional movement favors collective motion in the direction imposed by the environment. In contrast, model II imposes diffusion of cells along a principal axis of anisotropy and leads to dispersion of the cells. For both models, the cells respond positively to an increase of the field strength and their response saturates for infinitely large drives.

Note that apart from cell migration, the microenvironment plays an important role in the evolutionary dynamics (as a kind of selective pressure) of evolving cellular systems, like cancer ([Anderson 2006, Basanta 2008a]). It is evident that a profound understanding of microenvironmental effects could help not only to understand developmental processes but also to design novel therapies for diseases such as cancer.

In summary, a module-oriented modeling approach, like LGCA, hopefully contributes to an understanding of migration strategies which contribute to the astonishing phenomena of embryonic morphogenesis, immune defense, wound repair and cancer evolution. In a following chapter of this thesis, we use the insight gained from these models to design a computational model of *in vivo* tumor invasion.

CHAPTER 6

The impact of migration and proliferation

Contents

6.1	Introduction	**83**
6.2	The LGCA model	**84**
	6.2.1 LGCA dynamics	85
	6.2.2 Micro-dynamical equations	86
6.3	Results	**87**
	6.3.1 Simulations	87
	6.3.2 Mean-field analysis	88
	6.3.3 Traveling tumor front analysis	96
6.4	Summary	**98**

6.1 Introduction

In this chapter, we attempt to answer the question, what is the impact of tumor proliferation and migration on tumor invasion. To do that we introduce a simple LCGA model of tumor invasion consisting of interacting tumor cells and necrotic entities[1]. In particular, we are interested in the effect of proliferation and migration on the velocity of the tumor's invasive front. Our analysis aims at predicting the velocity of the traveling invasion front, which depends upon fluctuations that arise from the motion of the discrete cells at the front. By means of a mean-field approximation, we are able to derive a macroscopic partial differential equation (PDE) describing our system. This equation characterizes the spatio-temporal tumor expansion at the tissue level. Introducing a *cut-off* in the macroscopic mean-field description allows for a quantitative characterization of the traveling wavefront. We calculate analytically the front speed and we compare it with the values derived from simulations. Finally, our analysis enables us to express tumor spreading by known tumor cell features, such as cell motility and proliferation rate. Finally, we provide an analytical estimate of the front width and we demonstrate that it is proportional to the front speed.

[1]Necrotic entities correspond to sub-volumes of necrotic material

6.2 The LGCA model

We consider a lattice gas cellular automaton defined on a two-dimensional regular lattice $\mathcal{L} = L_1 \times L_2 \in \mathbb{Z}^2$, where L_1, L_2 are the lattice dimensions. Let b denote the coordination number of the lattice, that is $b = 4$ for a square lattice. Cells move on the discrete lattice with discrete velocities, i.e. they hop at discrete time steps from a given node to a neighboring one, as determined by the cell velocity. The set of velocities for the square lattice as considered here, is represented by the two-dimensional channel velocity vectors $\mathbf{c_1} = \begin{pmatrix} 1 \\ 0 \end{pmatrix}$, $\mathbf{c_2} = \begin{pmatrix} 0 \\ 1 \end{pmatrix}$, $\mathbf{c_3} = \begin{pmatrix} -1 \\ 0 \end{pmatrix}$, $\mathbf{c_4} = \begin{pmatrix} 0 \\ -1 \end{pmatrix}$, $\mathbf{c_5} = \begin{pmatrix} 0 \\ 0 \end{pmatrix}$,. In each of these channels, we consider an exclusion principle, i.e. we allow at most one cell per channel. We denote by $\tilde{b} = b + \beta$ the total number of channels per node which can be occupied simultaneously, where β is the number of channels with zero velocity (rest channels), here $\beta = 4$ [2]. In our LGCA, we represent healthy tissue by the empty channels and we model explicitly two cell "species", denoted by $\sigma \in \Sigma = \{C, N\}$: tumor cells (C) and necrotic (N) entities, respectively. We allow the movement of these populations in two different parallel [Dab 1991] lattices $\mathcal{L}_\sigma \sim \mathcal{L}$. We represent the channel occupancy by a Boolean random variable called *occupation number* $\eta_{\sigma,i}(\mathbf{r}, k) \in \{0, 1\}$, where $i = 1, ..., \tilde{b}$, $\sigma \in \Sigma$ for tumor cells and necrotic entities, $\mathbf{r} \in \mathbb{Z}^2$ the spatial variable and $k \in \mathbb{N}$ the time variable. The $\tilde{b}$-dimensional vector

$$\boldsymbol{\eta}_\sigma(\mathbf{r}, k) := \left(\eta_{\sigma,1}(\mathbf{r}, k), ..., \eta_{\sigma,\tilde{b}}(\mathbf{r}, k)\right) \in \mathcal{E}$$

is called *node configuration* and $\mathcal{E} = \{0, 1\}^{\tilde{b}}$ the automaton *state space*. *Node density* is the total number of cells present at a node $\mathbf{r}$ for a given species σ, and denoted by

$$n_\sigma(\mathbf{r}, k) := \sum_{i=1}^{\tilde{b}} \eta_{\sigma,i}(\mathbf{r}, k)$$

We can also define a *total node density* that is the sum of node densities over all species σ

$$n(\mathbf{r}, k) := \sum_{\sigma = C, N} \sum_{i=1}^{\tilde{b}} \eta_{\sigma,i}(\mathbf{r}, k)$$

The *global configuration* for the lattice of species σ is given by

$$\boldsymbol{\eta}_\sigma(k) := \{\boldsymbol{\eta}_\sigma(\mathbf{r}, k)\}_{\mathbf{r} \in \mathcal{L}}$$

---

[2] The value of the number of rest channels β is defined upon scaling of the model to a corresponding experiment or in vivo situation. Since the model is not representing any specific experiment, the choice of β remains arbitrary here and qualitatively identical results were obtained for tests with different choices.

6.2. The LGCA model

Finally, the global configuration is $\eta(k) := \{\boldsymbol{\eta}_\sigma(\mathbf{r}, k)\}_{\mathbf{r} \in \mathcal{L}, \sigma \in \Sigma}$ and the overall state space is $\{0, 1\}^{\tilde{b}|\Sigma|}$.

6.2.1 LGCA dynamics

Automaton dynamics arises from the repetition of three rules (operators): Propagation (P), reorientation (O) and cell reactions (R). The composition of the three operators R ∘ O ∘ P is applied independently at every node of the lattice at each time step resulting in the next configuration:

$$\eta_{\sigma,i}^{\text{R} \circ \text{O} \circ \text{P}}(\mathbf{r} + m_\sigma \mathbf{c}_i, k+1) = \mathcal{R}(\eta_{\sigma,i}(\mathbf{r}, k)).$$

In particular, the re-orientation and the propagation operators are related to cell motion, while the cell reactions operator controls the change of the local number of cells on a node. Propagation (P) and reorientation (O) have been discussed previously (Chapters 2 and 3). Therefore, we present in detail our cell reaction operator (R).

6.2.1.1 Cell reactions (R)

In this section, we define the interactions between the two cell species and among individuals of each species. Generally, the definition of these interactions is a difficult and ambitious task. For *in vivo* tumors the complexity of the interaction phenomena cannot be captured easily by computational models. In our model, we try to include the most important features of tumor growth and we attempt to approximate the cell interactions. In this study, an important modeling assumption is that we relate the free space to nutrient availability.

- **Tumor cells**: Here, two processes are taken into account: *mitosis* and *necrosis*.

  - *Mitosis* is the cell doubling process. We assume that tumor cells can divide only if they have just a few competitors on the node (less competition for nutrients), i.e. the node density of tumor cells $n_C(\mathbf{r}, k)$ should be lower than a threshold $\theta_M \in (0, \tilde{b})$. The fixed probability of mitosis r_M could potentially be a function of tumor node density.

  - *Necrosis* is the decay of tumor cells due to nutrient depletion. Analogous to the above, if the total node density exceeds $\theta_N \in [1, \tilde{b})$, then we assume that the nutrient consumption is critical and leads to tumor cell necrosis. The fixed necrosis probability could be a function of $n_C(\mathbf{r}, k)$ and can be defined in various ways following *in vivo* and *in vitro* observations.

Now, we define the new node density after the action of the reaction operators for the tumor cells:

Chapter 6. The impact of migration and proliferation

$$n_C^R(r,k) := \begin{cases} n_C(\mathbf{r},k) + 1, & \text{w. p. } r_M \text{ if } n_C(\mathbf{r},k) \leq \theta_M \\ n_C(\mathbf{r},k) - 1, & \text{w. p. } r_N \text{ if } n_C(\mathbf{r},k) \geq \theta_N \\ n_C(\mathbf{r},k), & \text{else,} \end{cases} \quad (6.1)$$

where w. p. denotes "with probability". It is easy to observe that tumor cells undergo a birth-death process with corresponding probabilities $r_M, r_N \in (0,1) \subset \mathbb{R}$.

- **Necrotic entities**: Necrotic entities are produced from tumor cells by the process of necrosis, i.e. due to nutrient depletion within a node. The new node density of the necrotic entities is given by

$$n_N^R(\mathbf{r},k) := \begin{cases} n_N(\mathbf{r},k) + 1, & \text{w. p. } r_N \text{ if } n_C(\mathbf{r},k) \geq \theta_N \wedge n_N(\mathbf{r},k) < \tilde{b}, \\ n_N(\mathbf{r},k), & \text{else.} \end{cases}$$
(6.2)

Hence necrotic entities undergo a birth process with probability r_N. Obviously, necrotic entities play a passive role in the evolution of the tumor. Finally, we note that once created necrotic entities do not move. Note that it is reasonable to assume that $\theta_M < \theta_N$.

6.2.2 Micro-dynamical equations

Following the above description of the automaton rules, we try to derive the micro-dynamical description of our LGCA. The post-reaction state $\eta_{\sigma,i}^R(\mathbf{r},k)$ after the application of the reaction operator is:

$$\eta_{\sigma,i}^R(\mathbf{r},k) = \mathcal{R}_{\sigma,i}(\eta_{C,i}(\mathbf{r},k), \eta_{N,i}(\mathbf{r},k)), \quad (6.3)$$

where $\mathcal{R}_{\sigma,i} : \mathcal{E} \to \mathcal{E}$. In particular, eq. (6.3) can be written as:

$$\eta_{C,i}^R = \eta_{C,i} + \xi_{C,i}\tilde{\eta}_{C,i}\Theta(\theta_M - n_C) - \xi_{N,i}\eta_{C,i}\Theta(n_C - 1 - \theta_N) \quad (6.4)$$
$$\eta_{N,i}^R = \eta_{N,i} + \xi_{N,i}\tilde{\eta}_{N,i}\Theta(n_C - \theta_N), \quad (6.5)$$

where we dropped the space and time dependence for simplicity and used the notation $\tilde{\eta}_{\sigma,i} = 1 - \eta_{\sigma,i}$. The $\xi_{\sigma,i}$ are random Boolean variables which represent the realization of a mitotic or a necrotic event, with $\sum_{i=1}^{\tilde{b}} \xi_{\sigma,i} = 1$, and the corresponding probabilities are $\mathbb{P}(\xi_{C,i} = 1) = r_M/\tilde{b}$ and $\mathbb{P}(\xi_{N,i} = 1) = r_N/\tilde{b}$. The Heaviside function $\Theta(\cdot)$ is defined as:

$$\Theta(x) = \begin{cases} 1, & \text{if } x \geq 0 \\ 0, & \text{else.} \end{cases}$$

The complete spatio-temporal automaton dynamics are described by the following *micro-dynamical difference equations*:

$$\eta_{\sigma,i}(\mathbf{r} + m_\sigma \mathbf{c_i}, k+1) - \eta_{\sigma,i}(\mathbf{r},k) = \eta_{\sigma,i}^{R \circ O}(\mathbf{r},k) - \eta_{\sigma,i}(\mathbf{r},k) = C_{\sigma,i}(\boldsymbol{\eta}_\sigma(\mathbf{r},k)) \quad (6.6)$$

6.3. Results

for $m_\sigma \in \mathbb{N}, \sigma \in \Sigma$ and $i = 1, ..., \tilde{b}$. The term $C_{\sigma,i}(\boldsymbol{\eta}_\sigma(\mathbf{r}, k))$ is called *collision operator* and takes the values $\{-1, 0, 1\}$. Details of the collision operator can be found in the Appendix C.

6.3 Results

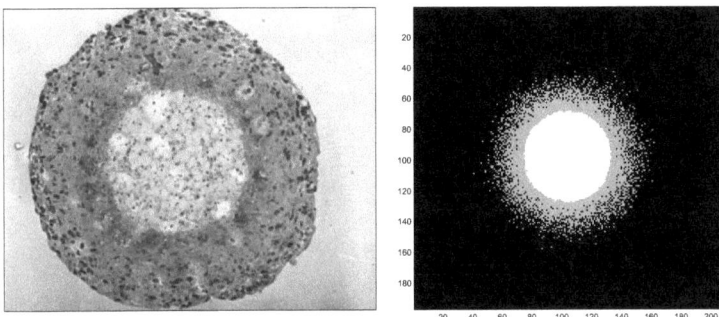

Figure 6.1: **Left**: Typical spatio-temporal pattern formation of *in vitro* tumors (reprinted with permission from Folkmann et al. [Folkman 1973]). One observes clearly the presence of a necrotic core and an outer rim of proliferative tumor cells. **Right**: A LGCA simulation exhibits a similar structure. In the simulation, tumor cells are depicted in grey, necrotic entities in white, and empty nodes in black. The comparison of the two figures is phenomenological, at the level of pattern formation, and not quantitative.

In this section, we focus on the numerical and the mathematical analysis of our tumor model. Firstly, we present simulation results with special emphasis on the system's pattern formation. In the following, by means of a cut-off mean-field analysis, we derive a macroscopic description of our LGCA. Finally, we analytically calculate the speed of the invasive front.

6.3.1 Simulations

In fig. 6.1, we observe the typical pattern formation of our system: the expansion of the the tumor front precedes the necrotic core because necrotic entities are created when the tumor cell density reaches the critical threshold θ_N. This pattern coincides with medical in vivo and in vitro observations, where typically tumors form a thin proliferating rim followed by a necrotic core [Folkman 1973].

For a further analysis a simplified, effectively one-dimensional geometry will be introduced. We employ two identical square lattices $\mathcal{L}_\sigma = L_1 \times L_2$ (L_1 represents

88 Chapter 6. The impact of migration and proliferation

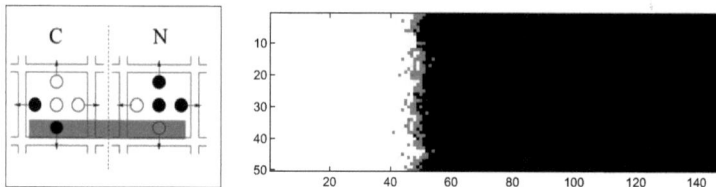

Figure 6.2: **Left**: Two corresponding nodes at position **r**, one from the tumor and the other from the necrotic lattice. The grey stripe denotes one chosen couple of channels. **Right**: Spatio-temporal pattern formation in the LGCA model. An invading two-dimensional tumor wavefront for $r_M = 0.2$ and $r_N = 0.7$. Tumor cells are depicted in grey, necrotic entities in white, and empty nodes in black.

the horizontal and L_2 the vertical axis of the lattice $\mathcal{L}_\sigma$, respectively), one for each cell species. The system is open at the right boundary of the L_1-axis and we have imposed zero-flux boundary conditions at the left boundary of the lattices. In the L_2-axis periodic boundary conditions were set. The initial condition (I.C.) is a fully occupied stripe of tumor cells at the beginning of the L_1-axis. The typical simulation time is 1500 time steps. The result of our simulations is a propagating two-dimensional front along the L_1-axis, mimicking "growth inside a tube" (fig. 6.2). The quasi one-dimensional setting has the following advantages:

- In order to study the traveling front, we reduce our two-dimensional system to one dimension, by averaging the concentration profile of each species along the L_2-axis, i.e. $n_x(k) = n(r_x, k) = \frac{1}{|L_2|}\sum_{r_y \in |L_2|} n(\mathbf{r}, k)$. Fig. 6.3 (left) shows that this simple model is able to create a traveling front that invades into the empty lattice nodes (healthy tissue).

- The front is well defined as the mean position of the foremost cells.

- The front profile relaxes to an almost steady state shape, which moves almost uniformly along the L_1-axis.

The goal is to study the front velocity and the front width. In the following sections, we provide the details of the front analysis. Finally, we observe that the tumor front evolves linearly in time, as shown in the right of fig. 6.3(right).

6.3.2 Mean-field analysis

In this section, we analyze the behavior of our tumor LGCA model. In particular, we derive a partial differential equation that corresponds to the automaton's macroscopic behavior, by means of a mean-field approximation. In the following, we introduce a cut-off in the mean-field description and we calculate the speed of the invasive front.

6.3. Results

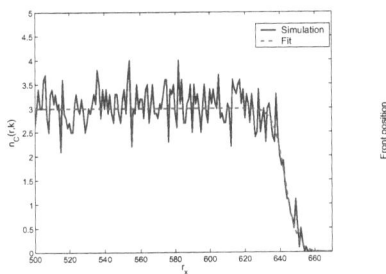
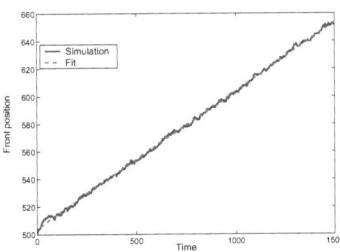

Figure 6.3: **Left**: Snapshot of the average concentration profile along the L_1-axis, i.e. $n_x(k) = n(r_x, k) = \frac{1}{|L_2|} \sum_{r_y \in |L_2|} n(\mathbf{r}, k)$. **Right**: Linear growth of the tumor front distance from its initial position, denoted as front position. The slope of the line defines the speed of the tumor invasion.

6.3.2.1 The non-linear Lattice Boltzmann equation

Let us define the single particle distribution functions which are the average values of the $\eta_{\sigma,i}$, i.e. the average channel occupation number

$$f_{\sigma,i}(\mathbf{r}, k) = \langle \eta_{\sigma,i}(\mathbf{r}, k) \rangle = \sum_{\boldsymbol{\eta}_\sigma} \eta_{\sigma,i}(\mathbf{r}, k) \mathbb{P}_k(\boldsymbol{\eta}_\sigma(\mathbf{r})),$$

where $f_{\sigma,i}(\mathbf{r}, k) \in [0,1]$, $i = 1, ..., \tilde{b}$. Note that the average $\langle ... \rangle$ is defined over an arbitrary node distribution $\mathbb{P}_k(\boldsymbol{\eta}_\sigma(\mathbf{r}))$ at time k.

Moreover, we define the *mean node density* as

$$\rho_\sigma(\mathbf{r}, k) = \langle n_\sigma(\mathbf{r}, k) \rangle = \sum_{i=1}^{\tilde{b}} f_{\sigma,i}(\mathbf{r}, k).$$

Applying the mean-field or Boltzmann approximation (Stoßzahlansatz), we can write down the *completely factorized* $\mathbb{P}_k$ distribution

$$\mathbb{P}_k(\boldsymbol{\eta}_\sigma(\mathbf{r})) = \prod_{i=1}^{\tilde{b}} \mathbb{P}_k(\eta_{\sigma,i}(\mathbf{r})) = \prod_{i=1}^{\tilde{b}} f_{\sigma,i}(\mathbf{r})^{\eta_{\sigma,i}(\mathbf{r})} (1 - f_{\sigma,i}(\mathbf{r}))^{1-\eta_{\sigma,i}(\mathbf{r})}. \quad (6.7)$$

The mean-field assumption discards all pair or higher on and off node correlations. Now one can derive from the micro-dynamical description (6.6) the mean-field approximation for our LGCA, which is called *non-linear Lattice Boltzmann Equation* (LBE)

$$f_{\sigma,i}(\mathbf{r} + m_\sigma \mathbf{c}_i, k + 1) - f_{\sigma,i}(\mathbf{r}, k) = \langle C_{\sigma,i}(\boldsymbol{\eta}(\mathbf{r}, k)) \rangle_{MF} = \tilde{C}_{\sigma,i}(\mathbf{f}(\mathbf{r}, k)), \quad (6.8)$$

where $\mathbf{f}(\mathbf{r}, k) = (\mathbf{f}_C, \mathbf{f}_N) = (f_{C,1}(\mathbf{r}, k), ..., f_{C,\tilde{b}}(\mathbf{r}, k), f_{N,1}(r, k), ..., f_{N,\tilde{b}}(\mathbf{r}, k))$ and $\tilde{C}_{\sigma,i} \in [-1, 1]$ is called *expected collision operator*. Considering that $m_C = 1$ and

$m_N = 0$, since the necrotic entities do not move, we can write the LBE for our model

$$f_{C,i}(\mathbf{r} + \mathbf{c}_i, k+1) - f_{C,i}(\mathbf{r}, k) = \frac{1}{b}\sum_{j=1}^{\tilde{b}}\langle \eta_{C,j}^{R}(\mathbf{r}, k)\rangle - f_{C,i}(\mathbf{r}, k)$$

$$= \frac{1}{b}F_C(\mathbf{f}_C, \mathbf{f}_N) \qquad (6.9)$$

$$f_{N,i}(\mathbf{r}, k+1) - f_{N,i}(\mathbf{r}, k) = \frac{1}{b}\sum_{j=1}^{\tilde{b}}\langle \eta_{N,j}^{R}(\mathbf{r}, k)\rangle - f_{N,i}(\mathbf{r}, k)$$

$$= \frac{1}{b}F_N(\mathbf{f}_C, \mathbf{f}_N). \qquad (6.10)$$

The F_σ terms can be easily calculated by applying the mean-field approximation to equations (6.4) and (6.5).

The steady states with homogeneous occupation, i.e. $f_{\sigma,i} = \bar{f}_\sigma$ can be determined numerically by solving the above expected collision operators when equal to zero (for details see the eqs. (C.2) and (C.3) in the Appendix), i.e.

$$\tilde{C}_{\sigma,i}(\bar{f}_C, \bar{f}_N) = 0 \Rightarrow (\bar{f}_C, \bar{f}_N) = (0, \alpha) \text{ or } (g(r_M, r_N), 1), \qquad (6.11)$$

where $\alpha \in [0, 1]$ is a free parameter and g a real function numerically determined, in dependence of the parameters r_M and r_N (fig. 6.4). The first fixed point represents the tumor-free situation. The second fixed point $(g, 1)$ corresponds to a well-defined necrotic core. The maximum number of tumor cells allowed to exist on a node that belongs to the necrotic core is equal to g.

For a given average number of cells per node $\bar{\rho}_\sigma$ the nonlinear LBE has a stationary and isotropic solution $\bar{f}_\sigma = \frac{\bar{\rho}_\sigma}{b} = \bar{u}_\sigma$. By eliminating the spatial effect, i.e by setting $m_\sigma = 0$, equations (6.9), (6.10) can be considered as a discretization of the ODEs

$$\frac{du_C}{dt} = \frac{1}{b}F_C(u_C, u_N) \qquad (6.12)$$

$$\frac{du_N}{dt} = \frac{1}{b}F_N(u_C, u_N), \qquad (6.13)$$

where F_C, F_N are rates of change of tumor cells and necrotic entities, respectively, expressed in terms of density per channel. The homogeneous, isotropic solutions $(\bar{f}_C, \bar{f}_N)$ of the LBE coincide with the fixed points of eqs. (6.12), (6.13).

6.3.2.2 Solutions of the linearized LBE

In order to gain insight into the behavior of the nonlinear LBE, we study small deviations from a steady state by linearizing around the homogeneous steady state solution. We define the *single particle distribution fluctuations*

$$\delta f_{\sigma,i}(\mathbf{r}, k) = f_{\sigma,i}(\mathbf{r}, k) - \bar{f}_\sigma. \qquad (6.14)$$

6.3. Results

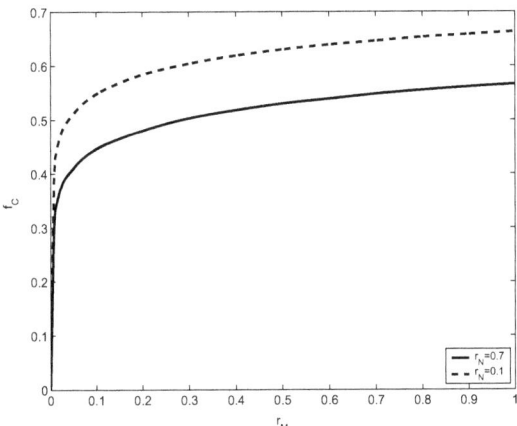

Figure 6.4: The second steady state solution of the tumor cells $\bar{f}_C = g(r_M, r_N)$ for different values of mitotic and necrotic probabilities. The mitotic and the necrotic thresholds are $\theta_M = 4$ and $\theta_N = 6$, respectively.

We linearize the nonlinear LBE around a steady state $(\bar{f}_C, \bar{f}_N)$ and construct the matrix $\boldsymbol{\Omega}^0$ with elements

$$\Omega^0_{ij} = \left.\frac{\partial \tilde{C}_{\sigma,i}(\mathbf{r},k)}{\partial \delta f_{\sigma,j}(\mathbf{r},k)}\right|_{\bar{f}_\sigma}, i,j = 1,...,\tilde{b}. \quad (6.15)$$

The matrix $\boldsymbol{\Omega}^0$ takes the form

$$\boldsymbol{\Omega}^0 = \begin{pmatrix} \frac{\partial \tilde{C}_{C,i}}{\partial \delta f_C} & | & \frac{\partial \tilde{C}_{C,i}}{\partial \delta f_N} \\ -- & | & -- \\ \frac{\partial \tilde{C}_{N,i}}{\partial \delta f_C} & | & \frac{\partial \tilde{C}_{N,i}}{\partial \delta f_N} \end{pmatrix}_{(f_C, f_N) = (\bar{f}_C, \bar{f}_N)}$$

with dimension $|\Sigma|\tilde{b} \times |\Sigma|\tilde{b}$ and the four block matrices have dimension $\tilde{b} \times \tilde{b}$, with $|\Sigma| = 2$ and $\tilde{b} = 8$. Hence, the linearized LBE is:

$$\delta f_{\sigma,i}(\mathbf{r} + m_\sigma \mathbf{c}_i, k+1) - \delta f_{\sigma,i}(\mathbf{r},k) = \sum_{i=1}^{\tilde{b}} \Omega^0_{ij} \delta f_{C,j}(\mathbf{r},k) + \sum_{i=\tilde{b}+1}^{2\tilde{b}} \Omega^0_{ij} \delta f_{N,j-\tilde{b}}(\mathbf{r},k). \quad (6.16)$$

Rearranging the terms of eq. (6.16) we obtain:

$$\delta f_{\sigma,i}(\mathbf{r}+m_\sigma \mathbf{c}_i, k+1) = \sum_{i=1}^{\tilde{b}} \boldsymbol{\Gamma}_{ij}(\mathbf{r},k) \delta f_{C,j}(\mathbf{r},k) + \sum_{i=\tilde{b}+1}^{2\tilde{b}} \boldsymbol{\Gamma}_{ij}(\mathbf{r},k) \delta f_{N,j-\tilde{b}}(\mathbf{r},k), \quad (6.17)$$

with
$$\Gamma_{ij} = \delta_{ij} + \Omega^0_{ij}, \qquad (6.18)$$
where the matrix $\Gamma = (\mathbf{I} + \mathbf{\Omega}^0)$ is called *Boltzmann propagator* which describes how small deviations from the $\bar{f}_\sigma$ evolve when the interaction operator R is applied on a node.

We choose as a linearization point the steady state (0,0), which represents the healthy tissue into which the tumor invades. Note that the behavior of the tumor at the tip of the invasive front is not influenced by the presence of the necrotic core, since this is developed far from the invasive zone, as shown in simulations. Moreover, the established necrotic region is related the second steady state $(g,1)$, where $g > 0$ (see fig. 6.4). Since the invasion happens at the boundary between tumor cells and healthy tissue (invasive point), the steady state (0,0) is the the only relevant for the study of tumor invasion dynamics.

After the linearization, the entries of the diagonal block matrices of the matrix $\mathbf{\Omega}^0$ are non-zero and all the others equal to zero. The propagator Γ takes the form

$$\Gamma = \mathbf{I} + \mathbf{\Omega}^0 = \begin{pmatrix} \omega_1 & | & 0 \\ -- & | & -- \\ 0 & | & \omega_4 \end{pmatrix},$$

where $\omega_1, \omega_4 \in \mathbb{R}_+$ are parameters to be defined below. The Boltzmann equation reads:

$$\delta f_{C,i}(\mathbf{r} + m_C \mathbf{c}_i, k+1) = \omega_1 \sum_{j=1}^{\tilde{b}} \delta f_{C,j}(\mathbf{r}, k) \qquad (6.19)$$

$$\delta f_{N,i}(\mathbf{r} + m_N \mathbf{c}_i, k+1) = \omega_4 \sum_{j=1}^{\tilde{b}} \delta f_{N,j}(\mathbf{r}, k), \qquad (6.20)$$

where $i = 1, ..., \tilde{b}$. Now, we insert the Fourier transform with wavenumber $\mathbf{q} = (q_1, q_2)$ of the corresponding Fourier mode:

$$\delta f_{\sigma,i}(\mathbf{r}, k) = A^k e^{i\langle \mathbf{q}, \mathbf{c}_i \rangle m_\sigma} \delta f_{\sigma,i}, \qquad (6.21)$$

where $\langle \cdot, \cdot \rangle$ is the inner product of two vectors. Then we obtain the following algebraic set of equations for the $\delta f_{\sigma,i}$'s:

$$\sum_{j=1}^{2\tilde{b}} \mathbf{M}_{ij} \delta f_{\sigma,i} = 0, i = 1, ..., \tilde{b}, \qquad (6.22)$$

where the $\mathbf{M}$ matrix is a block diagonal matrix with block matrices of 8×8 dimension, and its form is:

$$\mathbf{M} = \begin{pmatrix} -Ae^{i\langle \mathbf{q}, \mathbf{c}_i \rangle m_\sigma}\delta_{ij} + \omega_1 & | & 0 \\ -------- & | & -------- \\ 0 & | & -A\delta_{ij} + \omega_4 \end{pmatrix},$$

6.3. Results

A non-trivial solution exists if $det(\mathbf{M}) = 0$. Making explicit use of this condition, we obtain a 16th order polynomial equation for the damping coefficient A

$$A^{16} - 2A^{15}(\omega_1 cos(q_1) + \omega_1 cos(q_2) + 4\omega_4 + 2\omega_1) + 16A^{14}\omega_1\omega_4(cos(q_1) + cos(q_2) + 2) = 0, \quad (6.23)$$

and the solutions of A for the above discrete dispersion relation are:

$$A^{(1)}(\mathbf{q}) = 2\omega_1 cos(q_1) + 2\omega_1 cos(q_2) + 4\omega_1,$$
$$A^{(2)}(\mathbf{q}) = 8\omega_4,$$
$$A^{(j)}(\mathbf{q}) = 0, \text{for } j = 3, ..., 16.$$

The damping coefficients $A^{(1)}$ and $A^{(2)}$ depend on $\mathbf{q}$ and their value is different from zero. Then

$$\sum_{j=1}^{\tilde{b}} \mathbf{\Gamma}_{ij} = \sum_{j=1}^{\tilde{b}} (\delta_{ij} + \mathbf{\Omega}_{ij}^0) = \tilde{b}\omega_l \Leftrightarrow \omega_l = \frac{1}{\tilde{b}}(1 + \sum_{j=1}^{\tilde{b}} \mathbf{\Omega}_{ij}^0) \quad (6.24)$$

for $l = 1, 4$ and $i = 1, ..., \tilde{b}$. Following Boon et al. [Boon 1996] one realizes that

$$\sum_{j=1}^{\tilde{b}} \mathbf{\Omega}_{ij}^0 = \frac{1}{\tilde{b}} \frac{dF_\sigma(\bar{\mathbf{u}})}{d\bar{\mathbf{u}}} = -\kappa_\sigma, \quad (6.25)$$

i.e. κ_σ is the linearized phenomenological rate around the fixed point solution $\bar{u} = (\bar{u}_C, \bar{u}_N)$. Around the equilibrium point (0,0), we have $\kappa_C = -r_M$, $\kappa_N = 0$. Therefore:

$$\omega_1 = \frac{1}{\tilde{b}}(1 + r_M),$$
$$\omega_4 = \frac{1}{\tilde{b}}.$$

The damping coefficient $A^{(2)} = 1$ indicates a neutral stability against perturbations in the necrotic population, since necrotic entities, in the absence of tumor cells, do not proliferate, decay or migrate. Hence, the front propagation is solely driven by the linear instability $A^{(1)} > 1$ of the empty lattice against a perturbation with tumor cells.

For small wave numbers $|\mathbf{q}| \to 0$ and for infrequent cell divisions ($r_M \ll 1$) the damping coefficient $A^{(1)}$ can be expressed as the exponential of an equivalent continuous damping rate $z(\mathbf{q})$, i.e. $A^{(1)}(\mathbf{q}) = e^{z(\mathbf{q})}$ or

$$z(\mathbf{q}) = ln(A^{(1)}(\mathbf{q})) = ln(1 + r_M) - \frac{1}{\tilde{b}}|\mathbf{q}|^2 + O(|\mathbf{q}|^4). \quad (6.26)$$

For small mitotic rates $r_M \ll 1$, the diffusion is sufficiently rapid compared to reactions. Therefore, we can consider that reactions act as a perturbation of the diffusion process. Thus, for small mitotic rates the discrete rate law will closely approximate the continuous phenomenological rate [Tucci 2005]. The above equation shows that in this regime the dispersion relation is equivalent to that of the linearized reaction-diffusion equation:

$$\frac{\partial \delta u_C}{\partial t}(\mathbf{x},t) = D\nabla^2 \delta u_C(\mathbf{x},t) + \frac{1}{\tilde{b}}\frac{dF_\sigma(\bar{\mathbf{u}})}{d\bar{\mathbf{u}}}\delta u_C(\mathbf{x},t) \qquad (6.27)$$

with $D = 1/\tilde{b} = 1/8$ and $(\mathbf{x},t) \in \mathbb{R}^2 \times \mathbb{R}$ the continuous spatio-temporal variables. In this equation the field $\delta u_C(\mathbf{x},t) = u_C(\mathbf{x},t) - \bar{u}_C$ is the mass density fluctuation per channel on the lattice.
The continuous linearized reaction-diffusion equation (6.27) describes the time evolution of small perturbations around the fixed point $\bar{u}_C = 0$. Thus the fluctuations are $\delta u_C(\mathbf{x},t) = u_C(\mathbf{x},t)$, the linearized rate $\frac{1}{\tilde{b}}\frac{dF_\sigma(\bar{\mathbf{u}})}{d\bar{\mathbf{u}}} = -\kappa_C = r_M$ and the equation (6.27) multiplied by $\tilde{b}$ can be rewritten as:

$$\frac{\partial \rho_C}{\partial t}(\mathbf{x},t) = D\nabla^2 \rho_C(\mathbf{x},t) + r_M \rho_C(\mathbf{x},t) \qquad (6.28)$$

The above equation provides a macroscopic description of the tumor's spatio-temporal evolution around the fixed point $\rho_C = 0$ for small mitotic rates. When no tumor cells are present, the necrotic population remains unchanged, i.e. $\frac{\partial \rho_N}{\partial t}(\mathbf{x},t) = 0$.

6.3.2.3 Cut-off mean-field approximation

The spatio-temporal mean-field approximation (6.28) agrees qualitatively with the system's linearized macroscopic dynamics. However, it fails to provide satisfactory quantitative predictions because it neglects the correlations built by the local fluctuating dynamics. Studies on chemical fronts have shown that these fluctuations may significantly affect the propagation velocity of the wave front [Breuer 1994, Velikanov 1999].

In order to improve the mean-field approximation (here we characterize it as "naive"), we introduce the *cut-off mean-field approach* [Brunet 1997, Cohen 2005]. The idea is that the mean-field continuous equation (6.28) fails to describe the behavior of individual cells due to their strong fluctuations at the tip of the front [Breuer 1994]. Therefore, we derive the cut-off continuous approach which describes the system up to a threshold density δ of the order of magnitude of one cell, i.e. $\delta \sim \mathcal{O}(1/\tilde{b})$. Let's assume that the full non-linear reactive dynamics can be described by a term $F_C(\rho_C, \rho_N)$. Then, the fully non-linear cut-off MF equation reads

$$\partial_t \rho_C = D\nabla^2 \rho_C + F_C(\rho_C, \rho_N)\Theta(\rho - \delta), \qquad (6.29)$$

where $\Theta(\cdot)$ is a Heaviside function. Obviously, if we set $\delta = 0$ then the cut-off PDE will coincide with the naive mean-field approximation.

The cut-off macroscopic description (6.29) adds an extra fixed point, i.e. $\rho(x_i) = \{0, \delta, C\}$, $i = 0, \delta, C$ (by C we denote the maximum occupation defined by the

6.3. Results

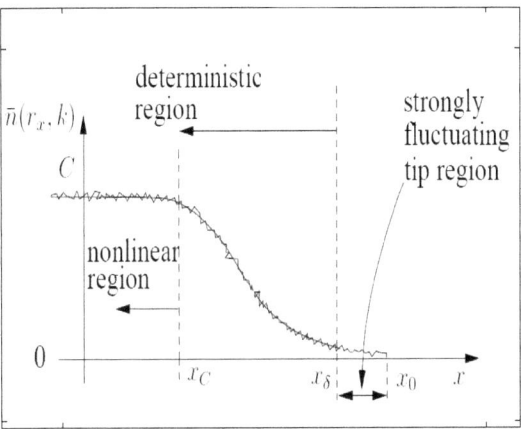

Figure 6.5: A sketch of the wavefront as shown in fig. 6.3 (left). We distinguish three regimes: (i) $x \in [x_\delta, x_0]$, where $0 < \rho(x) < \delta$: the region represents a highly fluctuating zone, where the cells perform a random walk with almost no proliferation, (ii) $x \in [x_C, x_\delta]$, where $\delta < \rho(x) < C$: this region is a result of non-linear proliferation and cell diffusion and (iii) $x \in [0, x_C]$, where $\rho(x) \simeq C$: this regime represents the bulk of the front (saturated lattice) where no significant changes are observed.

function $g(r_M, r_N)$ shown in fig. 6.4) which divide the front in the three following regions (fig. 6.5):

(i) $x \in [x_\delta, x_0]$, where $0 < \rho_C(x) < \delta$: this regime represents a highly fluctuating zone, where the cells perform a random walk with almost no proliferation.

(ii) $x \in [x_C, x_\delta]$, where $\delta < \rho_C(x) < C$: this region is a result of non-linear proliferation and cell migration.

(iii) $x \in [0, x_C]$, where $\rho_C(x) \simeq C$: this region represents the bulk of the front (saturated lattice) where no significant changes are observed.

In order to characterize the linearized tumor dynamics at the front, we modify the LBE for the tumor cells:

Chapter 6. The impact of migration and proliferation

$$f_{C,i}(\mathbf{r}+\mathbf{c}_i, k+1) - f_{C,i}(\mathbf{r}, k) = \sum_{j=1}^{\tilde{b}} (\frac{1}{\tilde{b}} - \delta_{ij}) f_{C,j}(\mathbf{r}, k) \quad (6.30)$$

$$+ \frac{1}{\tilde{b}} \sum_{j=1}^{\tilde{b}} \left[\langle \eta_{C,j}^{\mathrm{R}}(\mathbf{r}, k) \rangle - f_{C,j}(\mathbf{r}, k) \right] \Theta(\rho - \delta),$$

where the first summation of the rhs accounts for the reorientation dynamics and the second term is the reactive term of the LBE. Intuitively, the theta function "cuts off" the reaction term for local densities lower than the threshold δ. Therefore, for $\rho_C < \delta$ the cells are influenced only by the random walk dynamics. Moreover from eq. (6.30), we can easily deduce the nonlinear reaction term of eq. (6.29):

$$F_C(\rho_C, \rho_N) = \sum_{j=1}^{\tilde{b}} \left[\langle \eta_{C,j}^{\mathrm{R}}(\mathbf{r}, k) \rangle - f_{C,j}(\mathbf{r}, k) \right]. \quad (6.31)$$

6.3.3 Traveling tumor front analysis

In this subsection our goal is to analyze and characterize analytically the observed traveling front behavior. We consider that our system evolves in a "tube", as in fig. 6.2. Moreover, we make the following assumptions:

(A1) the isotropic evolution of the system allows for the dimension reduction of the analysis to one dimension,

(A2) the system evolves for asymptotically long times, and

(A3) the initial front is sufficiently steep.

Under the assumptions (A1)-(A3), we can consider that the front relaxes to a time invariant profile. Thus, assuming the translational invariance of the system along the front propagation axis L_1, we investigate the steady-state front solutions. The main observable is the average density profile along the axis L_1, i.e.

$$\rho_C(x,t) = \frac{1}{|L_2|} \int_0^{|L_2|} \rho_C(x, y, t) \, dy \in [0, \tilde{b}]. \quad (6.32)$$

Plugging the traveling front solution into eq. (6.28), $\rho_C(x,t) = U_C(x - vt)$, where $x \in L_1$ and v the front velocity, we obtain:

$$DU_C'' + vU_C' + \frac{d\tilde{F}_C}{dU_C}\bigg|_{U_C=0} = 0, \lim_{\xi \to -\infty} U_C = U_C^{max}, \lim_{\xi \to +\infty} U_C = 0, U_C' < 0, \quad (6.33)$$

in terms of the comoving coordinate $\xi = x - vt$ and the prime denotes the derivative with respect to the variable ξ. The term $\tilde{F}_C$ represents the reaction terms in the

6.3. Results

naive MF approximation expressed in terms of U_C and U_N. The front speed for the naive MF can be calculated following the classical methodology [Benguria 2004, Murray 2001], i.e.

$$v_n = 2\sqrt{Dr_M}. \qquad (6.34)$$

The above speed estimation overestimates the actual front speed found in the simulations. In particular, it is the maximum asymptotic value that the discrete front speed can acquire [Brunet 1997] (see also fig. 6.6).

The calculation of the front speed for cut-off MF approximation is more challenging. Following the results proposed by Brunet et al. [Brunet 1997], we can obtain an estimate for the cut-off front speed

$$v_c = 2\sqrt{Dr_M}\left(1 - \frac{K}{2\ln^2(\delta)}\right). \qquad (6.35)$$

The cut-off front speed estimation includes a correction factor $1 - \frac{K}{2\ln^2(\delta)}$, which allows for a better approximation of the actual front speed calculated from the LGCA simulations. The above equation provides a satisfactory description of the system up to the resolution of δ, i.e. to the order of one cell. A reasonable choice of the cut-off would be $\delta = 1/\tilde{b}$. The parameter K is fitted to match quantitatively the simulation results. Several studies have attempted to find an analytical estimate of K but till now remains an open problem [Brunet 2001]. The cut-off mean-field approximation is a heuristic-phenomenological approach which mimics the leading-order effect of finite population number fluctuations by introducing a cut-off in the MF equation. In fig. 6.6, we show a comparison of the front speed for varying proliferation rates r_M calculated by the naive MF and the cut-off MF against the front speed obtained from simulations. We observe that for an appropriate choice of K the cut-off MF predicts quantitatively the simulated front speed for all parameter values and a fixed choice of K.

Another important aspect of the invasive behavior is the width of the front. From fig. 6.5, we observe that exists the the front an inflection point $x = x^*$ where the derivatives $\partial_x^{(2n)}\rho_C|_{x=x^*} = 0$ and $n \in \mathbb{N}$. Typically, this inflection point is found at the middle of the front profile, i.e. $\rho_C(x^*) = \tilde{b}g/2$, where $g = \bar{f}_C$ refers to the bulk fixed point given in eq. (6.11) and it can be identified from fig. 6.4. The one dimensional, nonlinear cut-off MF approximation of the LGCA (6.29) at point $x = x^*$ reads:

$$\partial_t \rho_C|_{x=x^*} = [F(\rho_C, \rho_N)\Theta(\rho - \delta)]|_{x=x^*}. \qquad (6.36)$$

We transform coordinates into $\xi = x - vt$. Then eq. (6.36) is evaluated at the point $\xi^* = x^* - vt$ and it yields:

$$-vU_C'|_{\xi=\xi^*} = \tilde{F}(U_C, U_N)|_{\xi=\xi^*}, \qquad (6.37)$$

where the Heaviside function is equal to one, since $U_C(\xi^*) > \delta$. The width of the front is:

98 Chapter 6. The impact of migration and proliferation

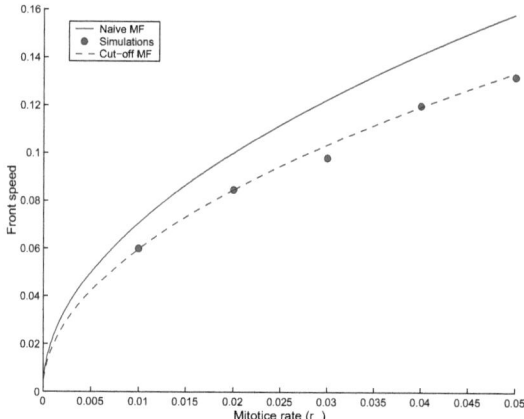

Figure 6.6: Comparison of the calculated front speed for the naive and the cut-off MF, i.e. v_n and v_c respectively, against simulations. We observe that the cut-off MF predicts closely the front speed calculated from the simulations for $K \simeq 1.7$.

$$W = -\frac{1}{U'_C|_{\xi=\xi^*}} = \frac{v}{\tilde{F}(U_C, U_N)|_{\xi=\xi^*}}. \quad (6.38)$$

As we observe the width of front is proportional to the front speed v. In order to calculate $\tilde{F}(U_C, U_N)|_{\xi=\xi^*}$, we use a uniform channel density $f_{C,i} = \rho_C(x^*)/\tilde{b} = g/2$ for $i = 1, ..., \tilde{b}$. Numerically the front width is estimated by fitting a straight line, tangential to the inflection point $\rho(x^*) = g/2$ such as in fig. 6.7 and the front width is approximated as the inverse slope of the fitted line.

Our analysis has shown that the front width depends directly on the front speed. As seen above in eq. (6.35), the front speed v is determined by fitting numerically the parameter K. Eq. (6.38) suggests that the same K, used for the calculation of v, allows for the prediction of the front width W. This result can be easily confirmed numerically.

6.4 Summary

Our focus was to establish a simple LGCA model of tumor invasion and to analyze the observed traveling front behavior. In the present study, we restrict our analysis to the characterization of the invading traveling front behavior in a homogeneous environment of two interacting populations of tumor cells and necrotic material. Via the cut-off mean-field analysis of the linearized discrete LBE, we derive a reaction-diffusion equation that describes our system macroscopically. This cut-off

6.4. Summary

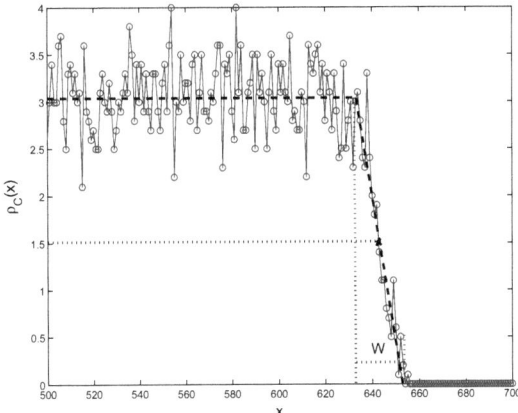

Figure 6.7: Numerically the front width is estimated by fitting a straight line, tangential to the inflection point $\rho_C(x^*) = \tilde{b}g/2$ and the front width W is approximated as the inverse slope of the fitted line. In this example, the simulation time is 1500 steps, the mitotic rate $r_M = 0.03$, necrotic rate $r_N = 0.7$, $|L_2| = 10$ and $g \simeq 0.38$ (it can be estimated from fig. 6.4). The black, dotted lines are indicating the methodology followed for the calculation of W.

R-D equation enables us to calculate accurately the speed of the tumor wavefronts.

The simulations show a "layer" formation of cancer and necrotic cells which is observed in numerous experiments [Folkman 1973, Bru 2003]. This behavior is a result of the tumor cell interactions. Mitosis creates an invasive tumor wavefront and the succeeding and increasing necrotic core follows the moving tumor border. Some experimental data suggest a linear growth kinetic for tumors [Bru 2003]. Our simulations show that the growth behavior of our model is consistent with these experimental observations. We predict the front velocity to scale with the square root of the product of rates for mitosis and migration. That means that we are able to calculate the tumor invasion speed by incorporating experimentally accessible parameters, like the mitotic rate and cell motility rate, respectively. The behavior of the front depends solely on the behavior of the tumor cells and not on the necrotic material. Another important aspect of the invasive behavior is the width of the front. It is of great interest the estimation of the length of tumor's invasive zone which coincides with the width of the traveling front. In particular, the invasive zone is of vital importance for the tumor since the majority of mitotic activity is concentrated within this zone. Thus, it is important to develop mathematical tools that allow the estimation of the tumor's front width. Here, we provide an estimate of front width which is proportional to the speed of the tumor front.

Chapter 6. The impact of migration and proliferation

A straightforward extension of the model is the consideration of the host's environment, as shown in [Hatzikirou 2008b]. Moreover, we can incorporate specific tumor cell dynamics, such as the so-called migration/proliferation dichotomy [Hatzikirou], and investigate the resulting tumor behavior. Finally, recent studies on the fractality of tumor surfaces [Bru 2003] have shown that tumors belong to a specific universality class of growth. The present study sets the basis for the analysis of more realistic and complicated models.

CHAPTER 7

Mechanisms of tumor invasion emergence

Contents

7.1	Introduction	101
7.2	Why not mutations?	105
7.3	The LGCA model	108
	7.3.1 The microenvironment: Oxygen concentration	108
	7.3.2 Cell dynamics	108
7.4	Simulations	111
7.5	Analysis	115
	7.5.1 Mean-field approximation	116
	7.5.2 Macroscopic dynamics	116
7.6	Summary	117
	7.6.1 Biological evidence	118
	7.6.2 Therapy	119

7.1 Introduction

Cancer progression can be described as a sequence of traits or phenotypes that cells have to acquire if a neoplasm (benign tumor) is to become an invasive and malignant cancer. A phenotype characterizes any kind of observed structure, function or behavior of a living cell. Hanahan and Weinberg [Hanahan 2000] have identified six possible types of cancer cell phenotypes, which are unlimited proliferative potential, environmental independence for growth, evasion of apoptosis, angiogenesis, high motility rates (invasion) and metastasis. It is widely believed that tumor cells change their phenotype due to mutations that are acquired during cancer progression.

Initially, mutations alter the proliferation control of the cells which leads to uncontrolled cell division [Hanahan 2000]. The transformed cells form a neoplastic lesion. The tumor can grow up to a size at which the diffusion-driven oxygen supply of the tumor becomes insufficient (hypoxia) to support further growth. Furthermore, the hypoxic environment and the high mutational rates of tumor cells, due to their damaged genetic material, may lead to the emergence of phenotypes characterized by anaerobic metabolism [Gatenby 2004], high motility or/and angiogenesis. These new attributes allow the tumor to grow further and at this stage metastases are often observed. However, the decisive difference between cells in benign and malignant tumors is the presence of cells with increased motility of the latter [Giese 2003]. We remind the reader that increased cell motility is closely associated with tumor invasive behavior, according to our definition in sec. 1.1. Since the emergence of a motile tumor cell population is so significant for tumor evolution towards malignancy, we focus here on the transition from the proliferative to the motile phenotype.

The principle question that we attempt to answer is which are the cellular mechanisms that promote the emergence of an invasive tumor phenotype at the expense of a proliferative one. The prevailing view concerning the emergence of invasive, or any other, phenotype is mutation-based. Random mutations of the appropriate combination of genes can switch the phenotype from a proliferative to an invasive one (or can trigger any other phenotypic change). However, the probability of "hitting" a specific combination of genes, that is related to a specific phenotypic change, is very low. For example, if we assume that changing a phenotype requires the mutation of N genes out of 30000 (approximate total number of genes in humans), then the probability of this phenotypic change is $p = 1/\binom{3 \times 10^4}{N} < 10^{-4}$. Moreover, let us assume that the occurrence of a phenotypic change follows a Poisson distribution with rate $\lambda = p/\Delta t$, where Δt is a time interval. Then, the time required for the occurrence of a such a phenotypic change is exponentially distributed and the expected time is $T = 1/\lambda > 10^4 \Delta t$, which can be very long. This fact contradicts with the fast evolution and the rapid adaptation of tumors like glioblastoma multiforme (GBM or simply glioma tumors). In particular, glioma tumors, even after an extensive resection, fully recur in less than six months. In the next section, we demonstrate that GBM recurrence is impossible to be explained by mutation-based theory. Therefore, there is a need for an alternative hypothesis.

Experiments with cultures of glioma cells [Giese 2003] have shown a relationship between migratory and proliferative behavior. Especially, cell motion and proliferation are mutually exclusive processes, since highly motile glioma cells tend to have lower proliferation rates, i.e. cells proliferate only when they do not move (resting phase). This phenomenon is known as migration/proliferation dichotomy (or *"Go or Grow"*) [Giese 1996b, Giese 1996a]. Biological evidence indicate that migratory and proliferative processes share common signalling pathways, suggesting a unique intracellular mechanism that regulates both behaviors [Giese 2003] ("Go or Grow" mechanism).

Based on the aforementioned biological observations, we propose an alternative mechanism that challenges the dominant hypothesis that mutations trigger the switch from a proliferative phenotype to an invasive one. We claim that the response of a microscopic intracellular mechanism, such as "Go or Grow", to oxygen shortage (hypoxia) maybe responsible for the transition from a highly proliferative

7.1. Introduction

to an invasive phenotype in a growing tumor (fig. 7.1).

Recently, several studies have investigated the influence of the migration/ proliferation dichotomy in tumor invasion. Athale et al. [Athale 2006] have proposed an agent-based model to test the effect of a potential regulatory network related to the "Go or Grow" mechanism in the emergence of invasive phenotypes. A lattice-based game theoretical approach [Mansury 2006], involving motile and proliferative populations, has been used to investigate the evolutionary dynamics of tumor growth. Finally, Fedotov et al. [Fedotov 2007] have been interested in the question what is the effect of the the "Go or Grow" mechanism on glioma cell diffusion, analyzed by means of a continuous random walk theory. However, none of the existing studies has tested our hypothesis: the response of "Go or Grow" mechanism to hypoxia may trigger the switch from a highly proliferative to an invasive phenotype.

To test and analyze our hypothesis, we use a lattice-gas cellular automaton (LGCA). Our LGCA models a phenotypically homogeneous, avascular tumor growing in a homogeneous oxygen concentration field. In our model, the cells accomplish two key functions: (i) cells execute an unbiased random walk and (ii) cell proliferation is influenced by the local oxygen concentration. The phenotype of the cells is controlled by a single model parameter, which represents the ratio between motility and proliferation rates. Varying the oxygen concentration allows us to identify the fittest tumor phenotype. The fitness quantifies the success of a phenotype under certain environmental constraints through its reproductive potential. In this chapter, fitness is characterized by the total number of tumor cells, i.e. offsprings, after a certain time interval. The simplicity of the model allows numerical and analytical investigations. In the following, we collect the most important biological implications of our study:

- We challenge the widely accepted theory that solely mutations are responsible for any phenotypic change involved in tumor progression. Our analysis suggests that the response of the microscopic "Go or Grow" mechanism to hypoxia, i.e. oxygen shortage, triggers the switch from a proliferative to a motile phenotype, and *vice versa* (fig. 7.1).

- For each level of oxygen supply, there exists a dominant (fittest) tumor cell phenotype that corresponds to a certain ratio of proliferation/migration rates.

- Finally, our model exhibits the non-intuitive behavior that invasive phenotypes, which show microscopically low proliferation and high motility, can produce more offsprings than phenotypes with much higher proliferation rates under certain micro-environmental circumstances.

In the next section, we answer the question why mutation-based explanations fail in the case of glioma tumor recurrence and why we need an alternative hypothesis. Then, we define a LGCA model that describes a growing tumor cell population. Furthermore, we introduce the assumptions of our model and we discuss the biological relevance of the model parameters. In section 7.3, we present the numerical

104   Chapter 7. Mechanisms of tumor invasion emergence

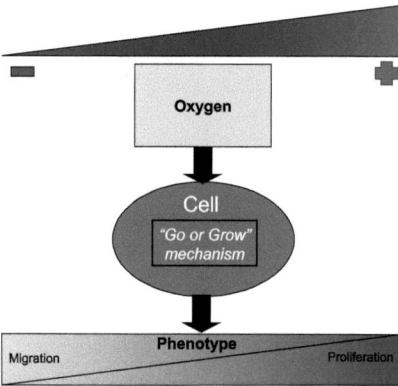

Figure 7.1: This flow diagram shows how the oxygen concentration influences the phenotype of a tumor cell. The "Go or Grow" mechanism respond to low oxygen supply with a motile phenotype. On the contrary, high oxygen level contributes to the occurrence of the proliferative phenotype.

7.2. Why not mutations?

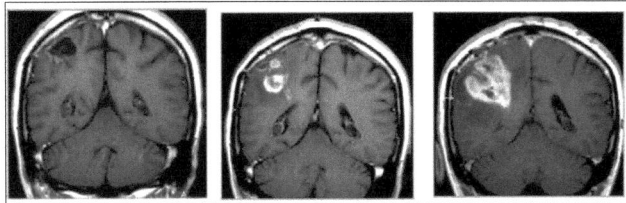

Figure 7.2: T1-weighted contrast-enhanced coronal MRI of recurrent glioblastoma. **Left**: T1-weighted contrast-enhanced coronal MRI section showing the resection cavity in the right parietal lobe – 3 months postoperatively. There is no evidence of tumor recurrence. **Middle**: corresponding MRI of the same patient, 6 months postoperatively, clearly showing tumor recurrence (=hyperintense or white mass). **Right**: control MRI, 9 months postoperatively, showing tumor extension beyond the previous resection cavity along the cerebral white matter.

results of the model simulations. Moreover, we determine the model's macroscopic behavior, through a multiscale Chapman-Enskog approach. Finally in section 7.4, we critically discuss the results and we elaborate clinical implications of our theoretical analysis.

7.2 Why not mutations?

In the introduction of this chapter, we claim that we challenge the prevailing view of mutation-driven cancer evolution. In this section, our goal is to demonstrate that the statement "Mutation-driven phenotypic changes are sufficient to explain the temporal evolution of all tumors" is false. We evidence the falseness of the above statement by the method of contradiction, i.e. by identifying an appropriate counterexample. A particularly interesting counter-paradigm is the recurrence of GBM tumors after extensive resection [Hatzikirou 2005], where a mutation-based theory fails to explain the speed of recurrence.

It is known that glioma tumors even after extensive resections, almost 99% of the tumor mass, fully recur in less than six months (fig. 7.2). Typically, a GBM tumor of has 10^9 cells (corresponds to a tumor diameter larger than 1cm) after surgery can be reduced up to mass proportional to $10^7 - 10^6$ cells. These remnant glioma cells are not resected because they have escaped far away from the bulk of the tumor (fig. 7.3). These cells typically belong to the invasive phenotype. According to the migration/proliferation dichotomy, highly motile cells are exhibiting low proliferation rates. The question that arises is how these invasive tumor cells are able to regenerate the initial tumor in such a sort time. In the following, we test if a mutation-driven phenotypic change is sufficient to support the recurrence the initial tumor within a timespan less than six months.

Chapter 7. Mechanisms of tumor invasion emergence

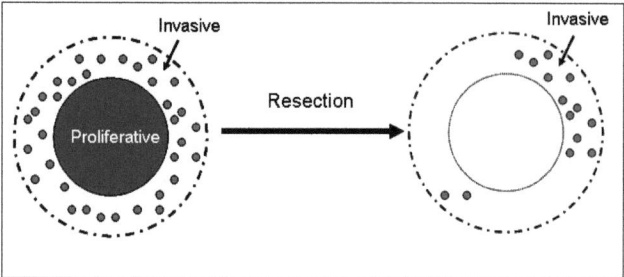

Figure 7.3: Sketch of pre- and postoperative of a glioblastoma tumor. In the pre-operative state a GBM tumor constitutes of an inner core of proliferative cells and an outer ring of invasive tumor cells (for simplification reasons the necrotic core is neglected). After a gross total resection, the main part of the tumor disappears but some cells of the invasive zone stay intact. These cells are assumed to be responsible for the tumor recurrence.

Firstly, we develop a simplified tumor model with two phenotypes, populations, a proliferative $\rho_p(t)$ and an invasive $\rho_i(t)$ one, respectively. The cells that belong to the invasive phenotype proliferate much less than the proliferative ones, i.e. the invasive proliferation rate is less than the proliferative one, $r_i < r_p$. We assume that mutations are responsible for the change of phenotypes, that the rate of change of phenotypes depends solely on mutations. The mechanism of mutation-driven phenotypic changes is readily depicted in fig. 7.4. In particular, let cells that belong to j^{th} population undergo mitosis with a constant rate r_j. Mutations occur in the newborn cells of population j and the proportion of cells that change to phenotype k is $m_j r_j \rho_j$, where $j \neq k$ and $j, k \in \{p, i\}$. It is plausible to consider that $m_i = m_p = m$ since there is no reason to assume that mutations could favor just one direction. The following system of equations describes the time evolution of the two phenotypes:

$$\frac{d\rho_p}{dt} = \overbrace{r_p \rho_p}^{proliferation} + \overbrace{mr_i \rho_i}^{gain\, i \to p} - \overbrace{mr_i \rho_p}^{loss\, p \to i} - \overbrace{d_p \rho_p}^{death} \quad (7.1)$$

$$\frac{d\rho_i}{dt} = \overbrace{r_i \rho_i}^{proliferation} + \overbrace{mr_p \rho_p}^{gain\, p \to i} - \overbrace{mr_i \rho_i}^{loss\, i \to p} - \overbrace{d_i \rho_i}^{death} . \quad (7.2)$$

The total tumor population at time t is given by the sum of the two phenotypes, i.e. $\rho(t) = \rho_p(t) + \rho_i(t)$. Please note, that the above model overestimates for long times the total tumor growth. However, this simple model suffices to demonstrate our hypothesis.

Now, let us assume that a GBM tumor of 10^9 cells is resected up to $99,9\%$, i.e. the post-operative glioma population counts up to 10^6 invasive tumor cells, i.e. the

7.2. Why not mutations?

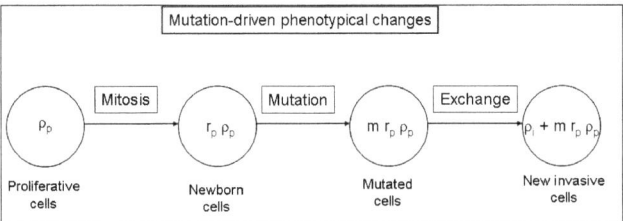

Figure 7.4: Diagrammatical representation of mutation-driven phenotypic changes (from the proliferative to the invasive phenotype). Cells that belong to the proliferative population (ρ_p) undergo mitosis and the newborn cells ($r_p \rho_p$) are subjected to mutation events. The mutation of the appropriate combination of genes leads to the change of phenotype in $m r_p \rho_p$ cells.

initial conditions after the resection are $(r_p(0), r_i(0)) = (0, 10^6)$, where $t = 0$ is the resection time. From the literature, we can determine the values of the mitotic and death rates for both phenotypes (Table 7.1).

Firstly, we assume that no phenotypic changes are required to achieve full recurrence, i.e. $m = 0$. In this limit case, we investigate if the invasive cells alone are able to reproduce the resected tumor. After 180 days the total population accounts for $\rho(t = 180) \propto 10^7$ cells, i.e. the recurrent tumor is two orders of magnitude smaller than the initial one. Therefore, we conclude that for the full tumor recurrence, in a time of 180 days, is required the contribution of proliferative cells, i.e. it is required the occurrence of phenotypic changes of tumor cells from invasive to proliferative phenotype ($m > 0$).

Now, the aim is to determine the minimal value of the phenotypic change rate m, based on mutations, that provides a fully recurrent tumor within six months $t = 180$ days. From the model, we calculate the minimal rate that allows for a full recurrence is $m_{min} = 10^{-3}$ changes/day. However, such a phenotypic change rate based on mutations is completely unrealistic. In particular, let us assume that the minimal requirement that a tumor cell switches phenotype is a point mutation. The human genome includes about 30000 genes, i.e. the probability to find the appropriate gene that switches an invasive phenotype to a proliferative on is $1/30000 \propto 10^{-4}$. In the literature, it is assumed that the maximum mutation rate is 0.01 mutation/gene/cell division [Spencer 2004]. Therefore, the maximum phenotypic change rate is estimated as $m_{lit} = \{rate\,of\,finding\,the\,right\,gene\} \times \{rate\,of\,mutations\,per\,gene\} \propto 10^{-4} \times 10^{-2} = 10^{-6}$, i.e. three orders of magnitude larger than m_{min}. Thus, this fact leads us to the conclusion that the switch between proliferative and invasive phenotype cannot be only mutation-driven. Consequently, we can assume that another mechanism should be responsible for this phenotypic change. In the following, we develop a LGCA to test our hypothesis that the response of the "Go or Grow" mechanism to hypoxia, i.e. oxygen shortage, triggers the switch from a proliferative

108 Chapter 7. Mechanisms of tumor invasion emergence

Parameter	Notation	Value	Reference.
Mitotic rate of proliferative phenotype	r_p	10^{-1} days^{-1}	[Spencer 2004]
Mitotic rate of invasive phenotype	r_i	10^{-2} days^{-1}	[Stein 2007]
Death rate of proliferative phenotype	d_p	10^{-2} days^{-1}	[Spencer 2004]
Death rate of invasive phenotype	d_i	10^{-3} days^{-1}	[Stein 2007]

Table 7.1: Parameters for the mutation model. The proliferative rates of the invasive phenotype found in [Stein 2007] refer to an *in vitro* culture of glioma cells.

to a motile phenotype, and *vice versa*.

7.3 The LGCA model

We consider a lattice gas cellular automaton defined on a two-dimensional regular lattice $\mathcal{L} = L_1 \times L_2 \subset \mathbb{Z}^2$, where L_1, L_2 are the lattice dimensions. Cells move on the discrete lattice with discrete velocities, i.e. they hop at discrete time steps from a given node to a neighboring one, as determined by the *single particle speed*. The set of velocities for the square lattice as considered here, is represented by the two-dimensional channel velocity vectors $\mathbf{c_1} = \begin{pmatrix} 1 \\ 0 \end{pmatrix}$, $\mathbf{c_2} = \begin{pmatrix} 0 \\ 1 \end{pmatrix}$, $\mathbf{c_3} = \begin{pmatrix} -1 \\ 0 \end{pmatrix}$, $\mathbf{c_4} = \begin{pmatrix} 0 \\ -1 \end{pmatrix}$, $\mathbf{c_5} = \begin{pmatrix} 0 \\ 0 \end{pmatrix}$,. In each of these channels, we consider an exclusion principle, i.e. we allow at most one particle per channel. We denote by $\tilde{b} = b + \beta$ the total number of channels per node which can be occupied simultaneously, where β is the number of channels with zero velocity, the rest channels.

7.3.1 The microenvironment: Oxygen concentration

In the pre-vascular stage of carcinogenesis the tumor has not yet acquired its own vasculature, the oxygen therefore has to diffuse from the surrounding blood vessels to the tumor. In this study, oxygen is assumed to be homogeneously distributed in the lattice and replenished to a given constant value each time step. We define the parameter $C \in [0, \tilde{b}]$ that represents the *maximum node occupancy* which depends on the oxygen availability. Therefore, we assume that the number of tumor cells supported in a node is correlated with the available oxygen on that node. We would like to stress that the parameter C plays a crucial role in the analysis of our model.

7.3.2 Cell dynamics

In our model, particle dynamics are identified by the LGCA rules. Automaton dynamics arise from the repetition of three rules (operators): Propagation (P), reorientation (O) and cell reactions (R). In particular, the reorientation and the propagation operators dictate with the cell transport and the cell reactions operator

7.3. The LGCA model

controls the change of the local number of cells on a node through a birth/death process. The cell dynamics are subjected to the following assumptions:

A1 Tumor cells move randomly (see Chapter 3).

A2 Cell mitotic and apoptotic rates depend on the local cell density.

A3 The "Go or Grow" mechanism influences the birth/death process (mitosis/apoptosis) of the tumor cells (for details see below).

Propagation (P) and reorientation (O) have been discussed previously (Chapters 2 and 3). Therefore, we present in detail only our cell reaction operator (R).

7.3.2.1 Cell reactions (R)

In our model, tumor cells are allowed to proliferate and to die. The migration/proliferation dichotomy plays a crucial role in the definition of these processes. Moreover, oxygen supply influences individual cell's death rate. In detail, we define a stochastic birth-death process for the tumor cells as follows:

- **Mitosis**: Abnormal proliferation is a principle characteristic of cancer cells. We assume that the mitosis rule depends on microscopic volume exclusion (A2), where $\tilde{b}$ is the node capacity. For the creation of a new cell on a node, the existence of at least one cell and at least one free channel are required, i.e.:

$$\mathcal{R}_i(\mathbf{r}, k) = \xi_i(\mathbf{r}, k)(1 - \eta_i(\mathbf{r}, k)), \qquad (7.3)$$

where $\xi_i(\mathbf{r}, k)$'s are random Boolean variables, with $\sum_{i=1}^{\tilde{b}} \xi_i(\mathbf{r}, k) = 1$, and the corresponding probabilities are:

$$\mathbb{P}(\xi_i(\mathbf{r}, k) = 1) = r_m \frac{\sum_{i=1}^{\tilde{b}} \eta_i(\mathbf{r}, k)}{\tilde{b}}. \qquad (7.4)$$

Here, r_m is the probability of occupying a channel, if at least one cell exists on the node. Growth dynamics that assume local volume exclusion are termed in the literature as *carrying capacity-limited* or *contact-inhibited* proliferation dynamics. The proliferation rate r_m cannot be deliberately chosen. Growth of tumor cells is dictated by the "Go or Grow" mechanism (A3), i.e. cells are allowed to proliferate only when they rest, i.e. when they are positioned on a rest channel. Therefore, we can easily derive that the proliferation rate is proportional to the number of rest channel β, i.e.

$$r_m = \beta \bar{r}_m, \qquad (7.5)$$

where $\bar{r}_m$ is a base cell proliferation rate.

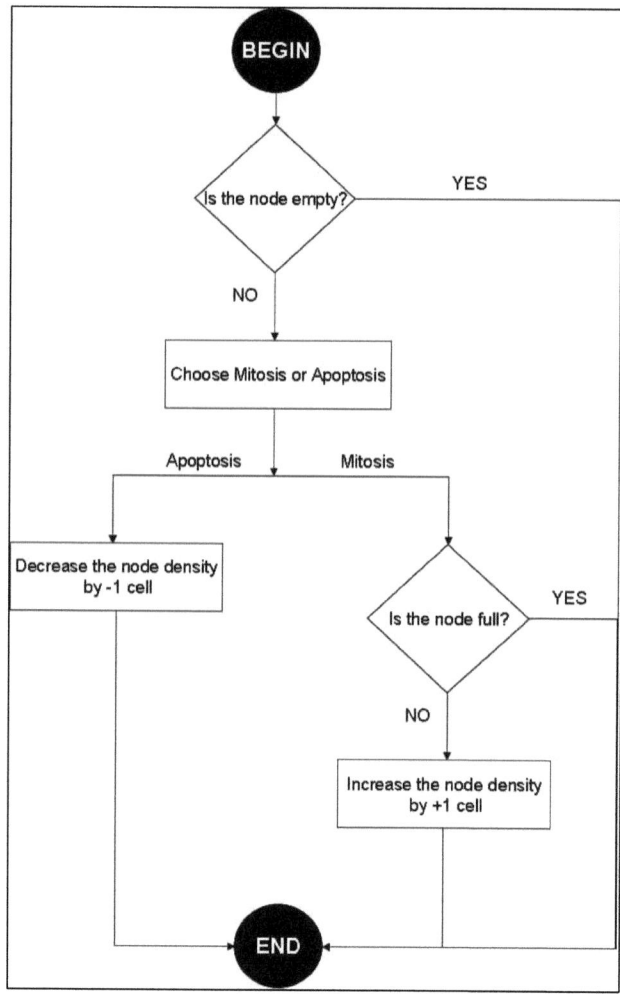

Figure 7.5: This flow-chart represents how the mitosis/apoptosis operator is modeled. The algorithm is applied for each node of the automaton. Initially, the algorithm checks if the node is empty. If not, cells can undergo mitosis or apoptosis depending on the probabilities (7.5) and (7.6), respectively. In the case of apoptosis the node looses a cell. On the other hand, mitosis is conditioned by the existence of at least one free channel. If the node is not full then the newborn cell is added to the node.

- **Apoptosis**: We have assumed that oxygen availability implies a maximum node occupancy C, i.e. the node oxygen supply cannot support more than C living tumor cells. Thus, we define a death rate for each tumor cell that ensures the existence of at most C cells per node (A2):

$$r_d = \frac{\tilde{b} - C}{\tilde{b}} \beta \bar{r}_m, \qquad (7.6)$$

where the factor $\frac{\tilde{b}-C}{\tilde{b}}$ is a dimensionless quantity. Note that, the death rate is monotonically increasing with respect to β, i.e. the number of rest channels (A3).

Fig. 7.5 shows schematically the implementation of the cell reactions operator.

Both mitotic and apoptotic cell rates depend on the number of rest channels β. In the next section, we demonstrate that the parameter β (number of rest channels) can be interpreted as the ratio of motility versus proliferation which characterizes the tumor cell phenotype. Moreover, cells with proliferative phenotypes, i.e. large β, exhibit larger rates of death and with invasive phenotypes lower rates of death (see (7.6)). This interpretation is consistent with experimental observations that indicate that invasive cells with *low propensity to proliferate also may be resistant to apoptosis* [Giese 2003].

7.4 Simulations

In this chapter, the principle question concerns the factors that trigger the switch from proliferative to invasive phenotypes. The "Go or Grow" mechanism imposes a relation between cell motility and proliferation. The crucial quantity is the total number of cells after a given period of time, since it quantifies the success of a phenotype at a given oxygen level, i.e. defines the fitness of the tumor in a given "environment". The control parameters are β, as a phenotype parameter and C, as an environmental parameter. The systematic variation of β allows us to identify the most successful phenotype in an environment characterized by C. In the following, we present a numerical analysis of our hypothesis and we provide the simulation results.

Firstly, we have simulated our LGCA model on a two-dimensional 500x500 lattice for 1000 time steps. In fig. 7.6, we show simulations for variations of the number of rest channels $\beta = 2, 4, 6$, for fixed maximum occupancy $C = \tilde{b}$ and for fixed base proliferation rate $\bar{r}_m = 0.05$. The initial condition is just a small disc. From the simulations, we conclude the following:

(O1) The pattern evolving in simulations from a localized initial occupation is an isotropically growing disc.

112　　　　　　　　　　Chapter 7. Mechanisms of tumor invasion emergence

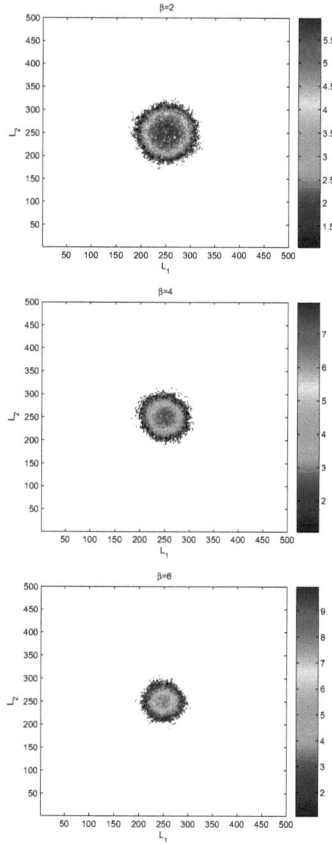

Figure 7.6: Typical pattern formation after 1000 time steps. For maximum occupancy $C = \tilde{b}$ and for a fixed proliferation rate $r_m = 0.05$, we vary the number of rest channels $\beta = 2, 4, 6$. We observe that, starting from an initial localized occupation, the motile populations (low β) expand faster than the proliferative ones. The colors encode the node density.

7.4. Simulations

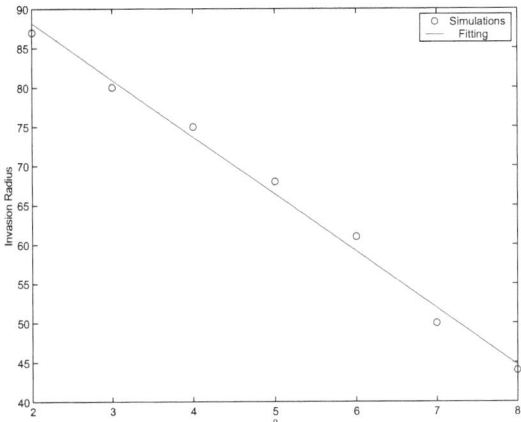

Figure 7.7: Infiltration radius against phenotype. For maximum occupancy $C = \bar{b}$ and for a fixed base proliferation rate $r_m = 0.05$, we show how the invasive radius varies by increasing the number of rest channels β, i.e. by making the tumor cells more proliferative. We observe a linear decrease of the infiltration radius as β increases.

(O2) Furthermore, simulations indicate a moving front along which the occupancy of the initially empty nodes is increasing from zero particles to the maximum occupancy C.

(O3) Increasing the number of rest channels, we observe that the disc radius decreases in size, since the cells become less motile. Moreover, the infiltration zone, i.e. the region between the periphery of the disc and the beginning of the core (maximum occupancy region), shrinks as β increases (fig. 7.7).

In order to get a deeper insight into the effect of the "Go or Grow" mechanism on tumor progression, we use a different simulation setup. We consider a "tube" (for details see subsec. 3.3.2.5), especially a 2000x10 lattice with periodic boundary condition on the L_2-axis, and a thin stripe of tumor cells as initial condition (fig. 7.8). A typical simulation time lasts for 2000 time steps. The result of our simulations is a propagating 2D front along the L_1-axis, mimicking a "growing tube". This setting has the following advantages:

- In order to study the traveling front, we project our system to 1D by averaging the concentration profile along the L_2-axis, i.e. $n(r_x, k) = \frac{1}{|L_2|} \sum_{r_y \in |L_2|} n(\mathbf{r}, k)$.

- The front is well-defined as the mean position of the foremost cells.

114   **Chapter 7. Mechanisms of tumor invasion emergence**

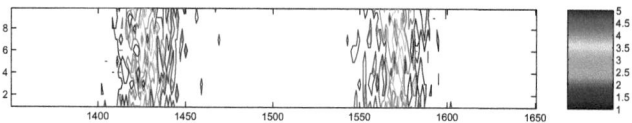

Figure 7.8: Typical simulation on a "tubular" lattice, i.e. with periodic boundary condition along the y-axis. The colors denote the node density. In the mid-section of the figure, the white part denotes nodes with maximum density.

- The diffusive dynamics of the front relaxes faster than the discoidal 2D evolution.

- The front profile relaxes to an almost steady state shape, which translates almost uniformly along the L_1-axis.

We simulate our system for different combinations of the parameters $\beta = \{1, 6\}$ and $C = 6, 7, 8$ (we fix the base proliferation rate to $\bar{r}_m = 0.05$). For these parameter combinations, we measure the total number of cells after the above-mentioned typical simulation time. The most interesting results are the following:

(O'1) The total number of cells evolves linearly in time.

(O'2) For each C iso-cline, i.e. *iso-nutrient curve*, we obtain a maximum value of the total number of cells for a unique β (fig. 7.9). This implies that the fittest phenotype is unique for a given oxygen supply C.

(O'3) Fig. 7.9 shows that under hypoxic conditions, i.e. after lowering the oxygen supply C, the fittest phenotype corresponds to the invasive one, i.e. higher motility and lower proliferation (low β).

(O'4) Moreover, we can observe in fig. 7.9 that a proliferative population (high β), can give rise to a lower number of offspring cells than populations of motile populations, i.e. low β, with lower proliferative rates. This result shows that the best strategy of cells in scarce resources environment is the faster exploration of new territories and the lowering of the proliferation rates.

Observations (O'2) and (O'3) confirm our main hypothesis that the response of the "Go or Grow" mechanism to a hypoxic environment favors the emergence of invasion.

At this point, we return to the case of glioma tumor recurrence. We have demonstrated that mutation-based phenotypic changes are sufficient to explain the speed of glioma recurrence. Therefore, an alternative hypothesis for the mechanisms that control the tumor phenotypic transitions is required. Now, we claim that the switch from a proliferative to a motile phenotype is controlled by the active interplay of "Go or Grow" mechanism and local oxygen supply. In particular, we claim that an

7.5. Analysis

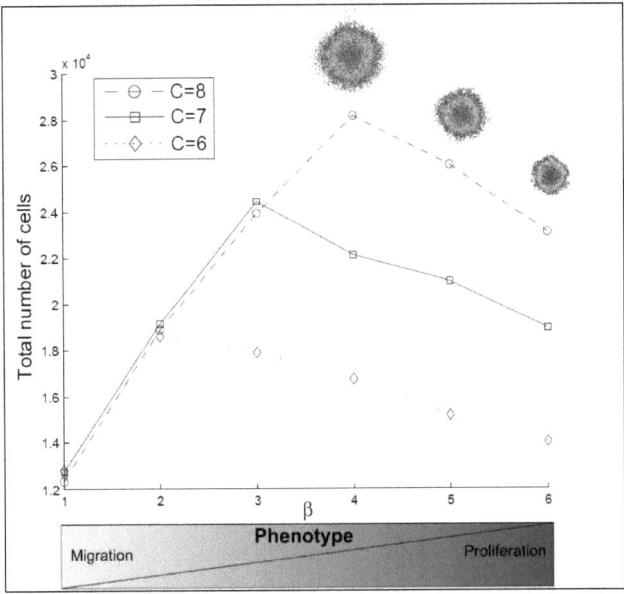

Figure 7.9: On the x-axis we vary the parameter β, which characterizes the tumor cell phenotype, ranging from motile populations (β small) to proliferative ones (β large). Each of the curves represents an iso-nutrient, i.e. the behavior of the population under the same oxygen availability. We observe that each iso-nutrient curve has a maximum point, which corresponds to the best fitted phenotype (β) in this specific environmental setting.

increase in the oxygen supply can accelerate the transition rates from an invasive to a proliferative phenotype (fig. 7.9). Especially in the case of the tumor bulk resection, the oxygen availability rises abruptly back to normoxic levels. Therefore, this fact can activate our proposed mechanism and consequently increase the transition rates towards a proliferative phenotype. This leads to an increase of the proportion of proliferative cells and the acceleration of the recurrent tumor's growth rates.

7.5 Analysis

In this section, we provide the mean-field approximation of our model. In particular, we derive the macroscopic diffusion coefficient and proliferative rate, which together define the phenotype of the population. The following mathematical analysis allows for the generalization of our numerical results.

7.5.1 Mean-field approximation

Our LGCA is governed by the following microdynamical equations:

$$\eta_i^R(\mathbf{r}, k) = \eta_i(\mathbf{r}, k) + \mathcal{R}_i(\mathbf{r}, k), \qquad (7.7)$$

$$\eta_i(\mathbf{r} + m\mathbf{c}_i, k + \tau) = \frac{1}{\tilde{b}} \sum_{j=1}^{\tilde{b}} \eta_j^R(\mathbf{r}, k). \qquad (7.8)$$

As seen in previous chapters (specially see Chapter 4), by averaging eqs. (7.7-7.8) and by using the MF approximation, we can obtain the lattice Boltzmann equation

$$f_i(\mathbf{r} + m\mathbf{c}_i, k + \tau) - f_i(\mathbf{r}, k) = \sum_{j=1}^{\tilde{b}} \Omega_{ij} f_j(\mathbf{r}, k) + \sum_{j=1}^{\tilde{b}} (\delta_{ij} + \Omega_{ij}) \tilde{\mathcal{R}}_j(\mathbf{r}, k), \qquad (7.9)$$

where the matrix $\Omega_{ij} = 1/\tilde{b} - \delta_{ij}$ is the transition matrix of the underlying shuffling process. Moreover, we assume that the mean-field reaction term is independent of the particle direction, i.e. $\tilde{\mathcal{R}}_i = F(\rho)/\tilde{b}$, where $F(\rho)$ is the mean-field cell reaction term for a single node. Using the mean-field approximation, we obtain the reaction term $\tilde{\mathcal{R}}_i$ is:

$$\tilde{\mathcal{R}}_i(\mathbf{r}, k) = r_m f_i(\mathbf{r}, k) \left(1 - \frac{r_d}{r_m} - f_i(\mathbf{r}, k)\right). \qquad (7.10)$$

7.5.2 Macroscopic dynamics

In order to derive a macroscopic description, we use the Chapman-Enskog methodology as described in Chapter 4. Here, we assume diffusive scaling as

$$\mathbf{x} = \varepsilon \mathbf{r} \text{ and } t = \varepsilon^2 k, \qquad (7.11)$$

where $(\mathbf{x}, t)$ are the continuous variables as $\varepsilon \to 0$. Furthermore, we assume an asymptotic expansion of f_i:

$$f_i = f_i^{(0)} + \varepsilon f_i^{(1)} + \varepsilon^2 f_i^{(2)} + \mathcal{O}(\varepsilon^3). \qquad (7.12)$$

Collecting the equal $\mathcal{O}(\varepsilon)$ terms, we can formally derive a spatio-temporal mean-field macroscopic approximation (see Chapter 4):

$$\partial_t \rho = \frac{m^2}{\tilde{b}\tau} \nabla^2 \rho + \frac{1}{\tau} F(\rho), \qquad (7.13)$$

where the term $F(\rho(\mathbf{r}, k)) = \sum_i^{\tilde{b}} \tilde{\mathcal{R}}_i(\mathbf{r}, k)$ is the macroscopic reaction law and using the definitions (7.5) and (7.6) we obtain:

$$F(\rho) = \frac{\beta \bar{r}_m}{\tilde{b}} \rho (C - \rho). \qquad (7.14)$$

Using the transformation

$$\rho \to \frac{\rho}{C}, \qquad (7.15)$$

we can translate (7.13) into an equation that belongs to the class of Fisher-Kolmogorov equations:

$$\partial_t \rho = \frac{m^2}{\tilde{b}\tau} \nabla^2 \rho + \frac{\beta \bar{r}_m C}{\tau \tilde{b}} \rho(1-\rho). \qquad (7.16)$$

With the above analysis, we have obtained an analytic approximation for the macroscopic evolution of our system. Eq. (7.16) is valid only for very small mitotic rates, i.e. $\bar{r}_m \ll 1$. The macroscopic proliferation and motility rates are identified by $\tilde{r}_m$ and D, respectively:

$$\tilde{r}_m = \frac{\beta \bar{r}_m C}{\tau \tilde{b}}, \qquad (7.17)$$

$$D = \frac{m^2}{(b+\beta)\tau}. \qquad (7.18)$$

We observe that the "Go or Grow" mechanism is evident in the above macroscopic coefficients since the proliferation rate (7.17) is monotonically increasing with respect to the number of rest channels (the more resting cells, the greater the proliferate rate) and the diffusive coefficient (7.18) monotonically decreasing (more resting cells induce reduced motility).

7.6 Summary

Here, we have investigated potential mechanisms that promote the progression from benign neoplasms to malignant invasive tumors characterized by high migration rates. In particular, we propose that the "Go or Grow" mechanism and oxygen shortage are sufficient to trigger the switch from a proliferative to an invasive phenotype in some cells. The need of an alternative hypothesis arises from the fact that mutation-driven tumor progression cannot explain the fast recurrence of glioma tumors after surgery. To test our new hypothesis, we set up a lattice-gas cellular automaton model. We represent the tumor phenotype by a single parameter β, corresponding to the ratio between proliferative and motile phenotypes. The tumor's microenvironment is represented by the oxygen concentration C, which is a crucial model parameter that influences the proliferative ability and the apoptotic rate of tumor cells. The parameter β plays a key role in the modeling of the "Go or Grow" mechanism, since its value is related to the proliferation rate of the cells. Interestingly, the apoptotic rate (7.6) is also related to the parameter β, implying that invasive cells with low propensity to proliferate also may be resistant to apoptosis [Giese 2003].

The analysis of our model indicates that indeed the response of the "Go or Grow" mechanism to oxygen shortage (hypoxia) may be responsible for the switch from a proliferative to an invasive phenotype, and vice versa. Our result provides an alternative explanation of the emergence of invasive phenotypes to the traditional hypothesis of mutation-driven tumor progression, to explain the emergence of invasive phenotypes. Our hypothesis can be considered as an example of phenotypic plasticity, i.e. microenvironmental effects on the emergence of tumor phenotypes [Merlo 2002, Anderson 2006, Quaranta 2008]. In particular, a correlation of a hypoxic environment with the emergence of invasive phenotypes has been only observed so far in game theoretical studies based on mutation-driven evolution, such as [Basanta 2008a, Basanta 2008b]. Moreover, we have shown that for each level of the oxygen supply C there exists a dominant (fittest) tumor cell phenotype that corresponds to a particular ratio of proliferation/ migration rates β_{max}. Therefore, our model allows for the prediction of the dominant phenotype given a particular oxygen concentration (fig. 7.9). Finally, our model exhibits the intriguing behavior that invasive phenotypes, which show microscopically low proliferation and high motility, may produce more offspring than phenotypes with much higher proliferation rates. This may explain, in part, the rapid expansion rates of many invasive tumors.

However, certain simplifications were made in our model:

- According to Hanahan et al. [Hanahan 2000], six tumor cell phenotypes have been identified, including the proliferative and the invasive one. *In vivo* other possible tumor phenotypes are commonly observed, such as glycolytic or angiogenetic ones. In this chapter, we focus on the transition from proliferative to motile phenotypes, since this transition is crucial for invasive tumors. However, we plan to extend our model by including more phenotypes and investigate the interplay between them.

- We do not assume interactions between tumor cells, such as cell-cell adhesion [Anderson 2005] or repulsion [Werbowetski 2004].

- The *in vivo* tumor microenvironment is far more complex, consisting of fibers, acidic substances, blood vessels etc. In our model, the oxygen concentration is considered as the only relevant environmental factor. The choice of the oxygen supply is based on biological evidence (see below) that links hypoxia with the emergence of invasive behavior.

7.6.1 Biological evidence

Hypoxia is a common feature of most cancers. Recently, several biological studies have linked hypoxia to the invasive behavior of tumors. In particular, it has been observed that hypoxia is responsible for down-regulation of cadherins, resulting is the disruption of cell-cell adhesive interactions, the promotion of invasive and metastatic behavior [Sullivan 2007] and the reduction of the proliferative activity [Daruwalla 2006]. These biological observations support our hypothesis that hypoxia triggers the switch from a proliferative to an invasive phenotype.

The dependence of the invasion/proliferation switch on the oxygen levels suggests that the regulatory network responsible for the control of mitosis and migration

7.6. Summary

may share common signalling pathways with the oxygen uptake network. Candidate molecular mechanisms that link hypoxia to invasive behavior have already been suggested. In particular, it has been found that in glioma cell lines hypoxia induces the expression of the c-Met protein which enhances glioma cell migration and invasiveness [Eckerich 2007]. The HIF protein has been recognized as a key molecule responsible for the hypoxia-induced tumor invasiveness [Fujiwara 2007].

Interestingly, in ecology we can identify a correlation of increased motility and species extinctions [Viswanathan 2008]. In particular, high motility strategies seem to confer a vital advantage in the limit of low densities – at the edge of extinction. Actually, empirical data indicate that some insects [Sisterson 2002] and fish [Lamine 2005] near starvation increase their movement intensity and diffusiveness in the search for food when compared to their foraging activity under normal conditions. During scarce environmental conditions, species presenting such behavior can have an adaptive advantage with respect to those which do not. As already observed above, it may determine what species do and do not perish in extreme conditions.

7.6.2 Therapy

Clinically, hypoxia correlates with an adverse prognosis and renders tumor cells more resistant to radiation and chemotherapy [Daruwalla 2006, Kizaka 2003]. Hence, improving tumor oxygenation may result in better treatment outcomes. This paradigm has been investigated in numerous clinical studies [Daruwalla 2006]. Our results point to a different therapeutic paradigm that could potentially be applied to many cancers. Certain brain tumors (gliomas) are a typical example for invasive and diffusive tumors. Gliomas not only proliferate but also actively invade the surrounding brain parenchyma. The surgical resection of these diffusive tumors will not result in a cure since the cancer cells have already invaded the surrounding healthy and functional brain tissue. This leads to recurrence of the tumor in all but a few cases. The prognosis for patients suffering from malignant gliomas is very poor. It has been suggested [Giese 2003] that invasive glioma cells are able to revert to a proliferative cellular program and *vice versa*, depending on the environmental stimuli. Our model suggests that re-activation of the proliferative program of invasive tumor cells by increasing oxygen tension (medical term for oxygenation) in the tumor will enhance the dominance of proliferative over the motile phenotypes. In theory the result will be a tumor characterized by more confined growth pattern and lower expansion speed. The analysis of our model showed that the switch from a motile to a proliferative phenotype will not lead to a faster growing neoplasm. Quite to the contrary, overall tumor growth will slow down. Hence, this therapeutical strategy may directly improve the patient's prognosis. It may also allow for more radical and therefore successful tumor resection.

CHAPTER 8
In vivo tumor invasion

Contents

8.1	Introduction	**121**
8.2	The algorithm	**123**
	8.2.1 Image segmentation	124
	8.2.2 Tumor modeling	127
	8.2.3 Shape analysis	130
	8.2.4 Evolutionary algorithm	131
8.3	Results	**134**
8.4	Summary	**136**

8.1 Introduction

In the previous chapters, we developed different models to shed light on the different aspects of tumor invasion. Here, we use the insight gained in the previous chapters to design a computational algorithm that is able to simulate *in vivo* tumor invasion. In particular, the objective of this chapter is to develop and implement an algorithm that optimizes the parameters of a lattice-gas cellular automaton model of tumor invasion, based on time-series medical data. These parameters may allow to reproduce clinically relevant tumor invasion patterns, providing a prediction of the tumor growth at a later time stage. This algorithm has been developed in the context of the master thesis of M. Tektonidis [Tektonidis 2008].

Here, we focus on brain tumors, especially on the most common and malignant brain tumor, the *Glioblastoma multiforme* (GBM). The World Health Organisation (WHO) distinguishes four grades of tumor malignancy: WHO grade I-IV [Hatzikirou 2005]. Glioblastoma multiforme WHO grade IV is the most frequent glioma and accounts for more than 50% of all primary brain tumors. Tumors of WHO grade II or III may evolve into GBMs, through a process termed as malignant progression. GBMs grow in three-dimensional, irregular patterns and infiltrate the surrounding brain tissue. The tumors frequently seem to grow along the fibers of the white matter or along vessels, i.e. appear to follow physical structures in the brain. The location of fiber tracks in the brain is provided by the data of *diffusion tensor imaging* (DTI). The GBM tumor type is suitable for the proposed algorithm of this chapter because of its rather deterministic evolution and its propensity to grow along the fiber tracks [Swanson 2002]. These properties should allow for a better prediction of GBM development than for other tumor types. Additionally, the gadolinium[1] enhancement of GBMs [Clark 1998] delivers MRI data of high tumor contrast, offering the possibility for a more accurate segmentation. Therefore, the above reasons make GBM an appropriate tumor paradigm that allows for mathematical modeling.

The algorithm consists of four modules. These modules interact with each other by exchanging data. The four modules of the algorithm are:

Image segmentation In MR images, the location and shape of a tumor can be easily recognized by the human eye. However, the boundary between tumor and healthy tissue is not clearly visible and is hard to be defined. The computer needs disambiguate information about the tumor's location and edge, in order to measure and analyze its shape. This is achieved by an image processing tool, the so-called *segmentation*. The segmentation process separates an image into uniform and homogeneous regions with respect to some characteristic such as grayscale value [Jähne 2004]. The result of the segmentation is a binary image divided into two regions, the *object of interest* (tumor) and the rest of the image, the *background*.

Shape analysis Once the image has been segmented and the tumor has been isolated from the rest of the image, its shape can be analyzed. In the first place, it might be useful to perform some corrections of the segmented image [Pratt 2007]. After the required corrections, the task consists of the measurement of reasonable *features* which are descriptive for the shape of the tumor. The measured features should provide information about the location, size and geometry of the tumor shape. These measurements can be used for the comparison of tumor shapes.

Tumor modeling The processed medical data can be used by a mathematical model which simulates the invasive growth of the GBM tumor. Here, we develop a glioma LGCA which is based on the models described in the previous chapters.

Evolutionary algorithm The task of the estimation of a biologically relevant parameter set for the LGCA model is carried out by an *evolutionary algorithm* (a well known example are the genetic algorithms). Evolutionary algorithms (EA) are evolving a population of possible solutions to the problem (i.e. parameter sets for the LGCA), until a satisfactory solution has been obtained.

---

[1] Gadolinium, or gadodiamide, provides greater contrast between normal tissue and abnormal tissue in the brain and body. Gadolinium looks clear like water and is non-radioactive. After it is injected into a vein, Gadolinium accumulates in the abnormal tissue that may be affecting the body or head. Gadolinium causes these abnormal areas to become very bright (enhanced) on the MRI. This makes it very easy to see. Gadolinium is then rapidly cleared from the body by the kidneys.

The solutions are evaluated by a *fitness function*, which assigns fitness values. Evolution occurs due to the creation of new generations of solutions, which are product of the recombination of solutions of the current population. Fitter solutions are more likely to get selected for reproduction, pass their properties onto the next generation [Mitchell 1996]. The problem of optimizing parameters for a model (LGCA), is expressed in terms of EAs as a problem of searching for better solutions in the *solution space*, a space with all possible solutions to the problem [Bentley 1999].

In the following section, we describe the overall design of the algorithm and the coupling of the different modules. Moreover, we present the details of the different modules. The result of this chapter is a *proof of principle* that the algorithm can identify the parameters through the images of a two-dimensional model-generated "tumor". Finally, we discuss the efficiency and the possible extensions of the algorithm.

8.2 The algorithm

Here, the implementation of the algorithm is restricted to a two-dimensional tumor development. Therefore, the algorithm requires two-dimensional data, which are obtained by the selection of a slice of the MRI and DTI volume data[2]. Usually, an axial plane of the brain is selected that is parallel to the plane of the machine and perpendicular to the longitudinal axis of the body. The selected slice should contain a representative slice of the tumor volume. Also, the LGCA model is defined and operates on a two-dimensional lattice, which is defined by the selected MRI slice.

The first step of the application is the segmentation of MR images. For each run of the algorithm, two MR images are required. Both images are slices of the brain MRI data containing a tumor, from the same patient at two different time points: *time point A* and *time point B*. The slices have to be selected at the same position and angle. The segmentation process locates the tumor area of each MR image and delivers a binary image where the tumor is isolated from the background. Necrotic material in the inner region of the tumor is considered as part of the tumor. The segmented image of *time point A* will be used as initial conditions for the LGCA, while the shape of the visualized tumor depicted in image of *time point B* is analyzed. The extracted features of the tumor shape will be used by the fitness function of the evolutionary algorithm.

The goal of the evolutionary algorithm is the search for a parameter set, which when fed into the LGCA with the tumor shape of *time point A* as initial conditions, leads to the generation of tumor shapes that are sufficiently similar to the shape of the *time point B*. The EA initializes a population with randomly generated individuals, where each individual represents a parameter set for the LGCA. The parameter sets are fed into the LGCA model, which simulates the tumor growth

---
[2]MRI and DTI are 3D data that consist of "slices", i.e. axial planes of the brain

using as initial conditions the segmented image of *time point A* and the information of the DTI data which influences tumor cell motion. The resulting image, which is characteristic for a given parameter set of the LGCA, visualizes the simulated tumor shape. The fitness function of the EA evaluates the relevance of the simulated tumor pattern against the real one of the MR image at *time point B*. The evaluation is expressed by a value, i.e. the *fitness value*, which is assigned to the corresponding individual. Then, the main loop of the EA starts. The EA picks individuals from the population which are destined to be parents. The selection proceeds according to the fitness value of the individuals, i.e. fitter individuals have a higher probability to be selected. The parent individuals are recombined and mutated by the genetic operators of the EA, forming offspring individuals, which are then placed into the population of the new generation. The new individuals are fed into the model and the fitness function evaluates the simulation result. This loop continues until the EA produces an individual which holds an acceptable fitness value. A fitness value is acceptable if it exceeds a given threshold. If an acceptable value is found, the algorithm terminates and delivers as result the fittest individual of the current population. The above described iteration is repeated until an acceptable parameter set is found or until a given number of iterations is achieved. An overview of the application is shown in fig. 8.1.

The delivered individual represents a parameter set which is characteristic for a specific tumor growth pattern of a particular patient. This parameter set can be used for the prediction of tumor patterns for future time points. In the following, we present in detail the different modules of the algorithm.

8.2.1 Image segmentation

Here, we describe the implemented segmentation process for MR images. A MR image represents a slice of the MRI volume data. We focus on brain MR images which contain a GBM. The tumor appears highlighted against the surrounding healthy tissue. For the specific problem, the segmentation process considers one object of interest, the tumor. The non-tumor area of the image is considered as background and is irrelevant for the process. Therefore, the decision process of the segmentation is classifying each pixel as a tumor or as a non-tumor pixel.

In MR images, a tumor may not appear homogeneous in terms of brightness values. Usually, regions with low brightness values are visible in the interior of a tumor, which corresponds to regions of necrotic material. However, we focus on the development of the tumor boundary. Typically, necrotic material does not influence the tumor boundary, as it is located in the core of the tumor. Therefore, at the segmentation process, necrotic material objects are not of interest, and all pixels which are included by the tumor boundary, are considered as tumor pixels.

The segmentation of MR images is a complex and difficult process. For instance, the contrast agents, such as gadolinium, may not be absorbed homogeneously, resulting in a highly inhomogeneous intensity value distribution of the tumor in the image. In such case, the image shows an overlap of the tumor and spatially close

8.2. The algorithm

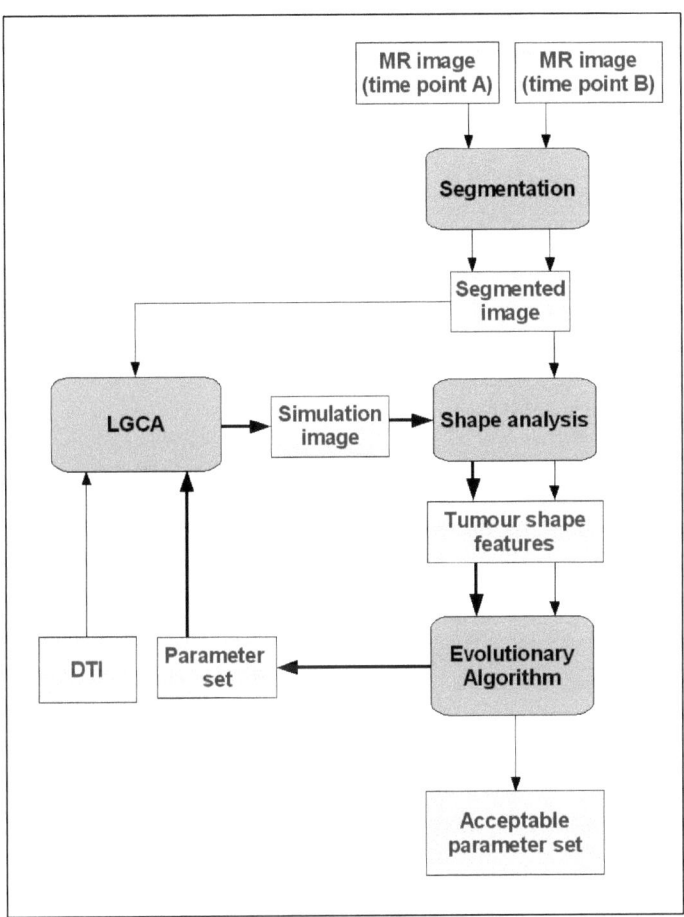

Figure 8.1: Illustration of the algorithm. The gray boxes represent the four main modules and the yellow boxes the data. The segmentation module operates only once for each MR image. The evolutionary algorithm, the lattice-gas cellular automaton simulation and the shape analysis are involved in the main loop (bold arrows) and exchange data. The algorithm terminates when the EA delivers an acceptable parameter set.

Chapter 8. *In vivo* tumor invasion

healthy tissue making tumor identification a cumbersome task. Experiments show that segmentation which was performed manually by medical doctors delivers results which vary significantly for the same MR image of a brain tumor. Generally, in the best case 90% of the GBM is visible in a MR image [Swanson 2002]. An incorrectly segmented image does not always imply the failure of the segmentation process, rather that the image itself does not contain enough information. Thus, the segmentation process provides an approximation of the tumor area rather than an accurate localization, since the segmentation has its limitations. Manual interactions of the user can decrease the probability of erroneous segmentation results.

Currently, there are no universally accepted segmentation methods which can be used for tumor MR images. A large number of different approaches has been developed [Clark 1998, Goldberg-Zimring 2005, Jiang 2004, Lundström 1997] but there will always be cases where a specific approach will not deliver a satisfactory result. Here, two algorithms have been implemented, the *thresholding* and the *region growing* algorithm, respectively. They represent two of the simplest segmentation approaches, however they provide satisfactory results for the tested images. Both are *semiautomatic*, as they require user interaction.

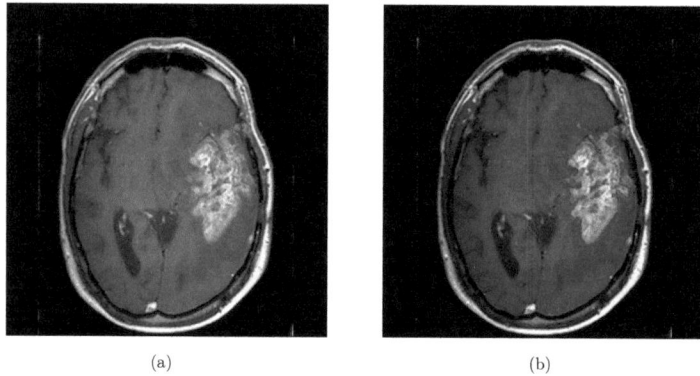

(a) (b)

Figure 8.2: Example of an image segmentation: **(a)** Original MR image, **(b)** tumour edge marked (red) on the original image.

The segmentation process will be performed twice; once it will be applied to the MR image of the tumor at *time point A* and once to the MR image of the tumor at *time point B*. The result of the first segmentation will be used for the LGCA model as initial conditions. In the following, the shape of the second segmented tumor will be analyzed, and the extracted features will be used for the evaluation of the LGCA simulations. Since an extended description of the image segmentation methods is out of the scope of this thesis, details can be found in [Tektonidis 2008].

8.2.2 Tumor modeling

In this section, we develop a simplified LGCA model of GBM growth. The main ingredients of this two-dimensional glioma model are the proliferative behavior of the tumor and the interaction with the fiber tracks (we use a slice-plane of the three dimensional DTI data). This model will be a combination of the models presented in Chapters 5 and 6 with small modifications.

In particular, the glioma model operates on a hexagonal lattice with 12-speeds (two-speeds per direction). The hexagonal lattice allows for higher order symmetry in comparison to the square lattice [Frisch 1987]. The addition of an extra speed per direction offers higher computational flexibility to our model [Wolf-Gladrow 2005]. In this model, we do not consider the evolution of the necrotic material, since they do not influence the evolution of the tumor's boundary (see Chapter 6). In the following, we introduce the *two-speed* LGCA model of tumor growth.

8.2.2.1 Prerequisites

We consider a two-dimensional hexagonal lattice $\mathcal{L} = L_1 \times L_2 \in \mathbb{Z}^2$. The coordination number is $b = 12$, since we assume two velocity channels per direction. The velocity channels are:

$$\mathbf{c}_0 = \begin{pmatrix} 0 \\ 0 \end{pmatrix}, \mathbf{c}_j = \begin{pmatrix} \cos \frac{j\pi}{3} \\ \sin \frac{j\pi}{3} \end{pmatrix} j = 1,...,6 \text{ and } \mathbf{c}_j = 2 \begin{pmatrix} \cos \frac{j\pi}{3} \\ \sin \frac{j\pi}{3} \end{pmatrix} j = 7,...,12 .$$

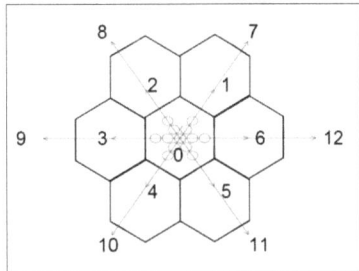

Figure 8.3: Simple velocity channels (1-6), double velocity channels (7-12), and rest channels (0) of a node in the hexagonal lattice of the *two-speed model* [Wolf-Gladrow 2005].

In this model, we fix the number of rest channels to $\beta = 6$. Thus, the total number of channels is $\tilde{b} = b + \beta = 18$.

In the *two-speed model*, automaton dynamics arises from the repetition of three rules (operators): Propagation (P), reorientation (O), cell reactions (R). Since the propagation operator has been described in detail previously (see Chapter 2), we do not repeat it here.

8.2.2.2 Cell reactions (R)

In this section we define the interactions between the two cell populations and among individuals of each population which is similar to the one used in Chapter 6. In particular, here we do not take into account the evolution of the necrotic material. For the sake of consistency, we repeat the essential parts of the definition of this operator.

Here, we try to include the most important features of tumor growth and we attempt to approximate the cell interactions. In this study, an important modeling assumption is that we relate the free space with nutrient availability. Two processes are taken into account: *mitosis* and *necrosis*.

- *Mitosis* is the cell doubling process. We assume that tumor cells can divide only if they just have a small number of competitors on the node (low competition for nutrients), i.e. the node density of tumor cells $n_C(\mathbf{r}, k)$ should be lower than a threshold $\theta_M \in (0, \tilde{b})$. The probability of mitosis r_M can potentially be a function of tumor node density.

- *Necrosis* is the decay of tumor cells due to nutrient depletion. Analogous to the above, if the total node density exceeds $\theta_N \in [1, \tilde{b})$, then we assume that the nutrient consumption is critical and leads to tumor cell necrosis. The necrosis probability is a function of $n_C(\mathbf{r}, k)$ and can be defined in various ways following *in vivo* and *in vitro* observations.

Now, we define the new node density after the action of the reaction operators for the tumor cells:

$$n_C^R(r,k) := \begin{cases} n_C(\mathbf{r}, k) + 1, & \text{w. p. } r_M \text{ if } n_C(\mathbf{r}, k) \leq \theta_M \\ n_C(\mathbf{r}, k) - 1, & \text{w. p. } r_N \text{ if } n_C(\mathbf{r}, k) \geq \theta_N \\ n_C(\mathbf{r}, k), & \text{else,} \end{cases} \quad (8.1)$$

where w. p. denotes "with probability". It is easy to observe that tumor cells undergo a birth-death process with corresponding probabilities $r_M, r_N \in (0, 1) \subset \mathbb{R}$.

8.2.2.3 Reorientation (O)

One central goal of our modeling effort is to include DTI data that in particular reflect the local anisotropy induced by the fiber track structure of the brain. The fiber tracks in the brain promote orientational changes for cell migration. In each point, a tensor (i.e. a matrix) informs the cells about the local orientation and strength of the anisotropy and proposes a principle (local) axis of movement defined by the vector field $\mathbf{E}(\mathbf{r})$, which is the principle eigenvector of the local diffusion tensor (see Chapter 5).

The re-orientation operator is responsible for the next time step velocity distribution of the cells. An alignment process is assumed along the local vector $\mathbf{E}(\mathbf{r})$,

8.2. The algorithm

provided by the processed DTI data. Here, we use an extended version of the reorientation operator used for alignment processes in Chapter 5. The transition probability of the node configuration $\boldsymbol{\eta}(\mathbf{r}, k)$ to $\boldsymbol{\eta}^O(\mathbf{r}, k)$ in an "oriented" environment is defined as:

$$\mathbb{P}(\boldsymbol{\eta} \to \boldsymbol{\eta}^O)(\mathbf{r}, k) = \frac{1}{Z} \exp\left(\alpha |\langle \mathbf{E}(\mathbf{r}), \mathbf{J}(\boldsymbol{\eta}^O(\mathbf{r}, k))\rangle|\right) \delta\big(n(\mathbf{r}, k), n^O(\mathbf{r}, k)\big). \quad (8.2)$$

where the post-collisional flux on a single node is

$$\mathbf{J}(\boldsymbol{\eta}^O(\mathbf{r}, k)) = \sum_{i=1}^{\tilde{b}} c_i \eta_i^O(\mathbf{r}, k) \quad . \quad (8.3)$$

The normalization factor Z is defined as:

$$Z(\boldsymbol{\eta}(\mathbf{r}, k)) = \sum_{\boldsymbol{\eta}^O(\mathbf{r}, k)} \exp\left(\alpha |\langle \mathbf{J}(\boldsymbol{\eta}^O(\mathbf{r}, k))\mathbf{E}(\mathbf{r})\rangle|\right) \quad . \quad (8.4)$$

Finally, for $\mathbf{E}(\mathbf{r}) = 0$ or $\alpha = 0$, we obtain a pure random walk process (Chapter 3).

8.2.2.4 Initial conditions

The lattice is initialized based on the segmented image of the MR data of *time point A*. This image defines the size $L_1 \times L_2$ of the hexagonal two-dimensional regular lattice and the initial distribution of cancer cells. The transformation of the Cartesian coordinates to the hexagonal ones is achieved by the use of the following equations:

$$u = x + \frac{y}{\sqrt{3}} \quad \text{and} \quad v = 2\frac{y}{\sqrt{3}} \quad , \quad (8.5)$$

where (u, v) are the hexagonal coordinate system and (x, y) the cartesian coordinates (see fig. 8.4). Then using eqs. (8.5) the size of the lattice is straightforward, i.e.

$$L_1 = X_{im} + \frac{Y_{im}}{\sqrt{3}} \quad \text{and} \quad L_2 = 2\frac{Y_{im}}{\sqrt{3}} \quad ,$$

where $X_{im} \times Y_{im}$ represents the size of the segmented image. Then, the mapping of the tumor pixels to the lattice is performed. Eqs. (8.5) are mapping each tumor pixel of the binary image to exactly one node of the hexagonal lattice. These nodes are initialized with the occupation of all their velocity channels, while the rest channels remain empty. Thus, their node density is $n(\mathbf{r}, 0) = 12, \forall \mathbf{r} \in \mathcal{L}_c \subset \mathcal{L}$, where $\mathcal{L}_c = \{\mathbf{r} \in \mathcal{L} | n(\mathbf{r}, 0) > 0\}$. After the initialization, the LGCA dynamics can be applied to the lattice.

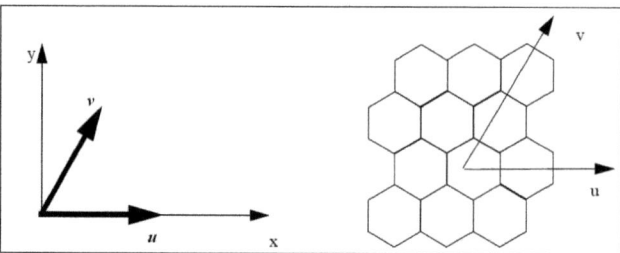

Figure 8.4: Hexagonal coordinate system described by the unit vectors **u** and **v**. The angle between **u** and **v** is 60°, while the corresponding angle in the orthogonal coordinate system between **x** and **y** is 90°.

8.2.3 Shape analysis

The task of this module is to analyze the tumor shape and extract a set of descriptive features. Several measurements of the shape will be used for the comparison of a specific tumor shape with other ones. The selection process of relevant features is connected with the fitness function of the evolutionary algorithm. The selected features should be representative of the size, location and shape of the tumor. The shape analysis module can be expressed mathematically as a function f:

$$f : \mathcal{I} \in \{0,1\}_{m \times n} \to \mathcal{F} \in \mathbb{R}^k, \qquad (8.6)$$

where $\mathcal{I}$ represents the set of two-dimensional binary images and $\mathcal{F}$ the set of the k-tuples of features, where each feature is described by a real number. Thus, the f maps a binary image (tumor pattern) to a measure $\mathcal{F}$ that characterizes this image.

The shape analysis module is applied both to binary images which are result of the segmentation process and to images generated by the simulation process of the lattice-gas cellular automaton. Regarding the MR images, only one of the two used images is analyzed, namely the one gathered at *time point B*. Images produced by the LGCA are analyzed and then the relevant features are compared with the ones of the MR image of *time point B*. In this way, each simulation result can be evaluated in terms of morphological similarity with the clinical image.

The obtained binary images can be measured, after a processing step, which involves the smoothening of the acquired images. The basic relevant measurements that can be performed are measurements for the area, perimeter, centroid, major and minor axis and orientation of the tumor. These values provide morphological information of the tumor shape. Afterwards, the shape descriptors, which combine these basic geometrical measurements can be extracted. Fig. 8.5 gives an example of the shape analysis for a circle, an ellipsoid and two simulations generated by the LGCA (figs. 8.5(c), 8.5(d)). The extracted features are then delivered to the evolutionary algorithm. A selection of them will be used by the fitness function

8.2. The algorithm

which evaluates each simulation result.

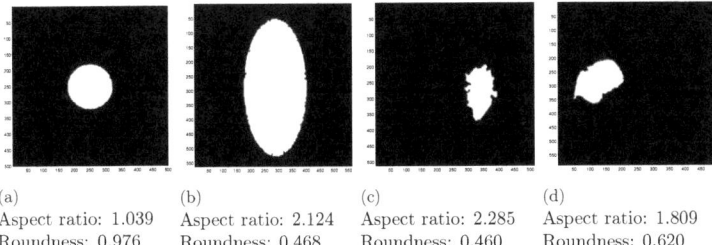

(a) Aspect ratio: 1.039 Roundness: 0.976
(b) Aspect ratio: 2.124 Roundness: 0.468
(c) Aspect ratio: 2.285 Roundness: 0.460
(d) Aspect ratio: 1.809 Roundness: 0.620

Figure 8.5: Shape analysis of four images; the aspect ratio and the roundness ratio are illustrated. **(a)** An approximately circular shape. **(b)** An ellipsoid. **(c)** An image produced by the LGCA; the segmented image of fig. 8.2 has been used as initial conditions for the LGCA. **(d)** Another simulation image.

8.2.4 Evolutionary algorithm

In the following, we describe the implemented evolutionary algorithm. First, we give some basic definitions related to EAs (see [Tektonidis 2008]):

- *(Potential) solution:* In our algorithm, a solution represents a parameter set for the LGCA model, i.e. any 5-tuple $(\theta_M, \theta_N, r_M, r_N, \alpha)$, with $\theta_M, \theta_N \in (0, \tilde{b}) \subset \mathbb{N}$, $r_M, r_N \in (0, 1) \subset \mathbb{R}$ and $\alpha \in (0, 10) \subset \mathbb{R}$, where $\tilde{b} = 18$ for the *two-speed model*.

- *Solution space:* The set of all possible solutions.

- *Individual:* The representation of a solution; e.g. *genetic algorithms* encode solutions into binary strings.

- *Search space:* The set of all individuals.

- *Population of individuals:* A subset of the search space.

The *problem* is to obtain a parameter set for the LGCA that is as representative as possible for the tumor development between *time point A* and *time point B* for a specific patient. The task of the evolutionary algorithm is to obtain such a parameter set by searching over the space of all possible solutions, using ideas of natural evolution. This search can also be considered as an optimization process of a given subset of solutions, which is usually created randomly.

The EA generates the initial population of individuals, which represents a subset of potential solutions to the given problem. The initial population will be used as the basis for the generation of new populations. The generation of new populations proceeds by the use of the evaluation process (*fitness function*) and the *genetic operators*.

132   Chapter 8. *In vivo* tumor invasion

8.2.4.1   Characterization of individuals

Our evolutionary algorithm does not use two separate spaces for the solution representation, i.e. the *search space* and the *solution space* are identical. The EA explicitly manipulates the solutions and does not use encoded versions of them, as it happens in *Genetic programming* and in *Evolutionary programming*. Therefore, the terms of *solution* and *individual*, are identical. While generating the initial population and manipulating the individuals, the EA is taking into account each parameter's value range in order to avoid the generation of individuals that do not belong to the search space.

8.2.4.2   Population size

During the run of the EA the population size remains constant. The size of the population is important as it affects the final result. The bigger the size of the population, the more individuals the algorithm will have available to generate a good solution. On the other hand, larger populations require more computation time and resources for the production of a new generation and its evaluation.

8.2.4.3   Fitness function

The fitness function of the EA serves for the evaluation of the solution quality. This evaluation involves the assessment of each individual with a fitness function and the assignment of a fitness value. This value will be used later to decide which individuals will have the chance of passing on their parameters to the next generation.

A reasonable selection of the fitness function in the problem of finding a relevant parameter set for the LGCA is a function that evaluates the similarity between the simulation result (using a specific parameter set) and the MR image of *time point B*. The simulated and the "real" tumor can be compared in terms of their shape characteristics, i.e. the extracted features (see section 6.4). The selection of the features which are compared is restricted to the ones that are the most descriptive for the tumor shape of the MR image (*time point B*). These selected features can be weighted with respect to their relevance. For some examples for the composition of the fitness function see in a later section.

As result of the comparison the fitness function delivers a value, and assigns it to the corresponding individual. The higher the fitness value of an individual is, the more clinically relevant is the parameter set for the given tumor growth between *time point A* and *time point B*. When a fitness value of a generated individual exceeds a given threshold, the parameter set which the individual represents, is characterized as *acceptable*. The EA is searching for an acceptable solution, rather than an optimal one. The selection of the fitness function has to be performed carefully in order to suit the given problem as much as possible. Generally, the quality of the result which the EA delivers, depends on the quality of the fitness function.

8.2. The algorithm

8.2.4.4 Genetic operators

For the reproduction process, the EA uses the genetic operators of genetic algorithms, particularly *selection*, *crossover* and *mutation*.

Selection

The selection operator chooses individuals from the population which will be involved in the reproduction process, i.e. the creation of new individuals. Selection is based on the fitness values of the individuals. Individuals with a high relative fitness have a higher probability to be selected, but candidates with relatively low fitness have a chance of passing on their parameters to the next generation too. If only the fittest candidates were selected for reproduction, diversity would be lost rapidly. With little diversity in the population, the EA may stop in a local maximum of the search space. An individual can be chosen more than once, and the operator terminates when the population of the new generation reaches the size of the current one.

- *Roulette-wheel selection.*
  The probability for each individual to get selected is equal to its relative fitness.

- *Tournament selection.*
  The operator picks randomly a number of candidates and the fittest of them is selected to be a parent.

The EA makes use of *elitism*. The fittest individual of the population is copied to the next generation and is not involved in the reproduction process. In this way, the best solution of each generation is maintained.

Crossover

The crossover operator uses the individuals chosen by the selection operator to create the individuals for the next generation. In our EA, two parent individuals are recombined and produce two offsprings which are placed into the new population. The implemented crossover operator is a 1-point crossover. Randomly, a single point is chosen, the same for both parents. Then the parents exchange segments after the crossover point, forming offspring. In this way, the new individuals inherit the properties (i.e. parameters) of their parents. The recombination of the parameters may imply that offspring individuals achieve a better fitness value, however this is not guaranteed.

Mutation

After the recombination of parent individuals, the mutation operator is applied to each offspring. Mutation randomly chooses a parameter of the offspring and modifies its value. The mutation rate is chosen randomly within a specific range (0%-15%) of the value range of the corresponding parameter. The mutation operator helps to maintain the diversity in the population and therefore helps the EA to avoid sticking at a search space's local maximum. A high mutation rate would cause a significant modification of an offspring, implying that the search performed by the EA would become essentially random. Thus, the implemented low mutation rate allows the offsprings to maintain the inherited characteristics of their parents, and simultaneously permit them minor modifications which may improve their fitness.

Termination criteria

As mentioned above, one criterion for the termination of the evolutionary algorithm is the generation of an acceptable individual (parameter set) with respect to the fitness function. The second termination condition occurs when a given number of generations have occurred. This implies that the EA has not found any acceptable solution. Usually, after a number of generations which depends on the population size, the population does not evolve significantly anymore. The result which the EA delivers in both cases is the individual with the highest fitness value.

8.3 Results

Now, we perform a characteristic experiment with the proposed algorithm. A vector field of DTI data of a real brain has been used. The dataset was provided by MeVis (Center for Medical Image Computing, Bremen). An image of *time point B* has been produced by the LGCA, using the image of *time point A* as initial conditions, the given vector field and a specific parameter set. Our application receives as input both images and the vector field, and delivers as output a parameter set.

In this example, we use the vector field derived from a DTI dataset and perform simulations with our above-defined glioma LGCA. Particularly, our application uses the following data:

- **Image of *time point A*** (Initial conditions for the LGCA): Image of fig. 8.6(a).

- **Image of *time point B***: Image of fig. 8.6(b); this image was produced by our glioma LGCA, with initial conditions defined by the use of fig. 8.6(a) and the parameter set $(\theta_M = 6, \theta_N = 9, r_M = 0.2, r_N = 0.8, \alpha = 5.0)$ for 30 simulation steps.

- **Vector field $E(r)$**: Vector field of DTI data (fig. 8.6(a)).

The fitness function of the evolutionary algorithm is given by:

$$C = \frac{1}{3}C_1 + \frac{1}{3}C_2 + \frac{1}{3}C_3 \quad ,$$

where:

$$C_1 = \begin{cases} \frac{A_s}{A_{im}}, & \text{if } A_s < A_{im} \\ \frac{A_{im}}{A_s}, & \text{else} \end{cases} \quad , \quad C_2 = \begin{cases} \frac{AR_s}{AR_{im}}, & \text{if } AR_s < AR_{im} \\ \frac{AR_{im}}{AR_s}, & \text{else} \end{cases}$$

$$\text{and} \quad C_3 = \begin{cases} \frac{O_s}{O_{im}}, & \text{if } O_s < O_{im} \\ \frac{O_{im}}{O_s}, & \text{else} \end{cases} \quad .$$

8.3. Results

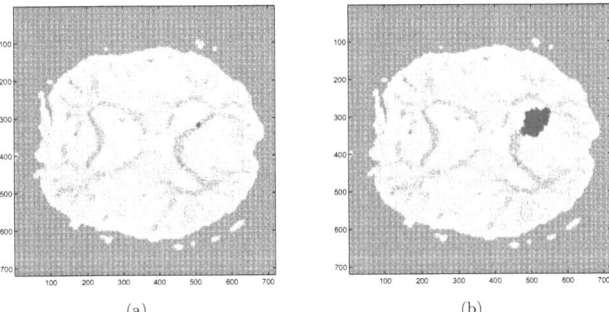

(a) (b)

Figure 8.6: Data of experiment: **(a)** Tumor of *time point A* mapped (red) on the vector field of the DTI data of a brain. The initial tumor has been placed on fiber tracks. **(b)** Tumor of *time point B* mapped on the vector field; note the simulated tumor growth along the fiber tracks.

	θ_M	θ_N	r_M	r_N	α
Optimal parameter set	6	9	0.2	0.8	5.0
Output parameter set	6	9	0.195	0.547	5.8
Error	0%	0%	0.5%	31.6%	16%

Table 8.1: Table with the output parameter set and the corresponding error.

$(A, AR, O) \in \mathcal{F}$ refer to the features *area*, *aspect ratio* and *orientation*, and the indices s and im refer to the image produced by the LGCA (simulation result) and the image of *time point B*, respectively. The area is descriptive for the size of the tumor shape, the aspect ratio and the orientation indicate the strength and the orientation of the elongation of the shape, which are controlled by the parameter α. The choice of weights is a plausible guess, since we are not aware of any sophisticated way of weighting the different image features in the fitness function. Note, that the necrotic core does not play any role in the fitness function, since it is irrelevant with the evolution of the tumor's boundary. Fig. 8.7 depicts some examples of simulation results and their evaluation by the defined fitness function.

The size of the individual population has been chosen as 100 cells. The EA initializes randomly the population and generates at every iteration step a new population. After 50 generations, the EA delivers the output parameter set (Table 8.3). The evolution of the maximum and average fitness of the population is illustrated in fig. 8.8.

The application delivers for this experiment an acceptable parameter set. Besides the high error for r_N the error for the other four parameters is relatively low.

For experiments that were performed but not presented here, the application made in most cases a satisfactory estimation of the four parameters r_M, r_N, θ_M and

Chapter 8. In vivo tumor invasion

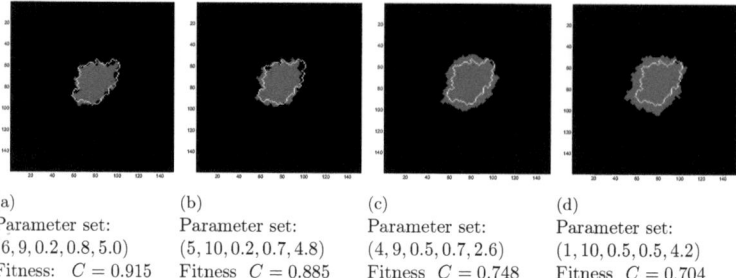

(a)
Parameter set:
$(6, 9, 0.2, 0.8, 5.0)$
Fitness: $C = 0.915$

(b)
Parameter set:
$(5, 10, 0.2, 0.7, 4.8)$
Fitness $C = 0.885$

(c)
Parameter set:
$(4, 9, 0.5, 0.7, 2.6)$
Fitness $C = 0.748$

(d)
Parameter set:
$(1, 10, 0.5, 0.5, 4.2)$
Fitness $C = 0.704$

Figure 8.7: Fitness function applied to some simulation results; the red color indicates the simulated tumor growth and the white line indicates the boundary of the tumor shape (image of *time point B*) (**a**) A simulation using the optimal parameter set achieves a high fitness value. (**b**) A similar parameter set to the optimal one, achieves a high fitness value. (**c-d**) Parameter sets that differ significantly from the optimal one achieve lower fitness values.

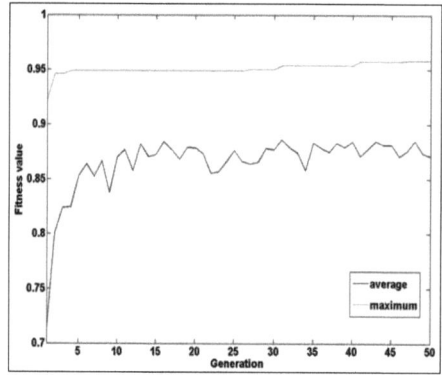

Figure 8.8: Evolution of the maximum and average fitness of the population. The EA terminates at generation 50 and delivers the fittest individual (fitness value 0.957).

α. The delivered necrosis rate parameter r_N has shown a high error. The reason for the failing of the implemented algorithm to estimate r_N is discussed in summary.

8.4 Summary

In this chapter, the focus was on designing a computational algorithm that would be able to predict *in vivo* tumor invasion. In particular, the main objective was

8.4. Summary

to obtain a parameter set for a cellular automaton model of tumor growth. The process of obtaining such a parameter set was interpreted as an optimization process, performed by the implementation of an evolutionary algorithm (EA). We focused on the development of Glioblastoma multiforme (GBM), the most common and malignant brain tumor. The propensity of GBMs to grow along fiber tracks of the brain makes the use of DTI data imperative, since these data provide information about the fiber track structure. Thus, the basic components of the optimization process are a set of medical data (MRI and DTI) and a lattice-gas cellular automaton (LGCA) of tumor growth. The proposed model is a composition of the tumor models proposed in the previous chapters of this thesis.

The algorithm delivered satisfactory parameter sets for the performed experiments, however the estimation of the parameter for the necrotic rate showed relatively high errors. These high errors occurred, due to the fact that necrotic material is irrelevant for the tumor's boundary development. Necrotic material, that is typically found in the core of a tumor, was neither considered in the segmentation process of MR images, nor in the visualization of the LGCA simulation result.

The main limitation of the proposed algorithm lies in the restriction to two-dimensional tumor development. The spatial development of tumors is three-dimensional. Hence, the two-dimensional LGCA model cannot produce clinically relevant tumor patterns, since it ignores the three-dimensional spatial development of tumors. Thus, for our numerical experiments, we used as input for the application artificially simulated images instead of real MR images. The use of artificial images offered us the possibility to verify the relevance of the developed algorithm at the two-dimensional level (proof of principle). The proposed method can be used as a basis for the development of a three-dimensional algorithm, which is able to deliver clinically relevant results. In this way, the developed an algorithm that could provide a potential tool for *in vivo* tumor invasion prediction.

CHAPTER 9

Discussion

Contents

9.1	Summary of the thesis	**139**
9.2	Can statistical mechanics help us to understand tumors?	**141**
9.3	Discussion and outlook	**142**
	9.3.1 Mathematical analysis of LGCA	143
	9.3.2 Tumor invasion modeling	143

9.1 Summary of the thesis

This thesis has been focused on the study of tumor invasion by means of mathematical modeling. The main processes involved in tumor invasion are related to tumor cell migration, cell-environment interactions and tumor cell proliferation. In order to better understand tumor invasion dynamics, we have used mathematical tools, motivated from statistical mechanics, that allow for modeling cellular processes and to analyze the emergent macroscopic behavior. Individual-based models, and especially CA, are well-suited for this task.

In this thesis, we use a special type of CA models, the so-called lattice-gas cellular automata [Deutsch 2005], which facilitate analytical investigations allowing for deeper insight into the modeled phenomena. The first part of the thesis is dedicated to the mathematical analysis of LGCA for basic biological processes, such as cell migration and proliferation. In particular, in Chapter 2 we introduce the LGCA nomenclature and important definitions. Then we present the main ingredients of discrete kinetic theory, i.e. the derivation of LBE as a mean-field manifestation of the LGCA microdynamics.

In Chapter 3, we have focused on the investigation of the most elementary cell migration mechanism, i.e. random cell motion. We have defined a LGCA model of random walk dynamics and we have derived the motility rate for (i) a single tagged particle and for (ii) a cell population. In the first case, we have calculated the Green-Kubo formula and we have calculated the single cell diffusion coefficient. In the cell population case, we have derived a macroscopic derscription from the definition of the LGCA. In particular, we have distinguished two different scalings, the parabolic and the hyperbolic, and we have derived to limit macroscopic equations, the diffusion and the telegraph equation, respectively.

Chapter 9. Discussion

In Chapter 4, we have investigated a LGCA cell growth process. Here we have focused on the application of a mean-field approximation under the assumptions of (i) a well-stirred system and (ii) a spatially distributed system. In the first case, we have derived an ODE that describes the temporal evolution of the total cell population. In the case of the spatially distributed system, we have provided the macroscopic dynamics by means of three different scaling methods, the Chapman-Enskog, the Coupled Map Lattice and the Fourier space methods. All three methods provide different macroscopic equations and they are valid for different regimes of the LGCA's parameters. Finally, we compare the well-stirred and spatially distributed system assumptions in terms of important macroscopic observables, such as per capita growth rate.

The second part of the thesis deals with the dynamics of tumor invasion. In particular, we have identified four basic questions concerning tumor invasion:

(Q1) *What is the impact of environment on tumor cell migration?*

In Chapter 5, we introduce a LGCA as a microscopic model of cell migration together with a mathematical description of different tumor environments. We study the impact of the various tumor environments (such as extracellular matrix) on tumor cell migration by estimating the tumor cell dispersion speed for a given environment.

(Q2) *How does tumor proliferation and migration influence a tumor's invasive behavior?*

In Chapter 6, we study the effect of tumor cell proliferation and migration on the tumor's invasive behavior by developing a simplified LGCA model of tumor growth. In particular, we derive the corresponding macroscopic dynamics and we calculate the tumor's invasion speed in terms of tumor cell proliferation and migration rates. Additionally, we find that the width of the invasive zone is proportional to the invasion speed.

(Q3) *Which are the mechanisms of tumor invasion emergence?*

The mechanisms for the emergence of tumor invasion in the course of cancer progression are investigated in Chapter 7. We conclude that the response of a microscopic intracellular mechanism (migration/proliferation dichotomy) to oxygen shortage, i.e. hypoxia, maybe responsible for the transition from a benign (proliferative) to a malignant (invasive) tumor.

(Q4) *How can we compute* in vivo *tumor invasion?*

In order to predict the spatio-temporal *in vivo* tumor invasion, we need to design appropriate computational algorithms. The main ingredients are: data, model and parameter evaluation. In Chapter 8, we propose an evolutionary algorithm that estimates the parameters of a tumor growth LGCA model based on time-series of patient medical data (in particular Magnetic Resonance and Diffusion Tensor Imaging data). These parameters may allow to reproduce clinically relevant tumor growth scenarios for a specific patient, providing a prediction of the tumor growth at a later time stage.

9.2 Can statistical mechanics help us to understand tumors?

In this thesis, our basic modeling (LGCA) and mathematical (discrete kinetic theory) tools have been drawn from the field of statistical mechanics (or statistical physics). Therefore, we are confronted with the question if statistical mechanics can help us to understand the biological properties of cell populations. In particular, we should answer how plausible is the use of statistical mechanics tools in the field of tumor biology? We remind the reader that throughout this thesis, tumors have been viewed as interacting multi-cellular systems.

Statistical mechanics is the physical theory that describes the connection between the micro-dynamics of multi-particle systems to their macroscopic observables. Traditionally, statistical mechanics provides the connection between microscopic motion of individual atoms of matter and macroscopically observable properties such as temperature, pressure, entropy, free energy or other thermodynamical variables. An extremely significant step in the history of statistical physics was the development of the kinetic theory by Boltzmann. The great innovation was that a coarse-grained description was introduced which kept the relevant information for the macroscopic dynamics but neglected the irrelevant microscopic details. The celebrated Boltzmann equation revolutionized statistical physics, since the mean dynamics of an ensemble of particles was enough to describe the full system, i.e. a gas. Finally, classical statistical mechanics are based on the following fundamental assumptions:

- Particles are identical.

- Microdynamics are based on particle collisions.

- Particle systems are close to the thermodynamic limit, i.e. their ensemble number in a volume element of space can be considered infinite (Avogadro number).

- Particle collisions are subjected to conservation laws.

- Particle collisions ensure microreversibility (detailed balance condition). A direct consequence of microreversibility and the thermodynamic limit assumption is the ergodicity of particle dynamics, i.e. statistical measures (such as averages) of particle ensemble can be described by equilibrium probability distributions.

The above assumptions allow for a satisfactory description of many physical systems. But the question that arises is what happens in biological systems? Is any of the above assumptions is valid? Until now, the only established connection of statistical mechanics is with molecular biology [Blossey 2006]. The study of molecular structures, like DNA, or the motion of molecular intracellular motors offer a fertile ground for the application of methods found in statistical mechanics.

But, what happens with the study of multi-cellular systems, such as tissues, organs or *in vitro* cell cultures. Can we apply tools from statistical mechanics for the study of such systems? On a first glance the answer is no! The fundamental assumptions of statistical mechanics fail in such systems. In particular:

- In contrast with particles, cell populations can be heterogeneous, because for instance cells can be differentiated.

- Cell interactions are quite different from particle collision and sometimes far more complicated.

- In some multi-cellular systems the "thermodynamic" limit assumption is not valid, i.e. the number of cells in a volume element can be relatively low.

- Cells are not subject to conservation laws. For instance, cell proliferation breaks the mass conservation law.

- Finally, the detailed balance condition is not valid in multi-cellular systems due to, for example, cell interactions with their environment (as shown in Chapter 5). Therefore, the concept of ergodicity is not valid.

Therefore, is our parallelization of multi-cellular systems with their physical counterpart, i.e. multi-particle systems, is completely misleading?

Fortunately, statistical mechanics are not restricted to the classical statistical physics of the 19th century. The mathematics of stochastic processes improved the classical tools of statistical mechanics and recently has been created a new field of non-equilibrium statistical mechanics [Zwanzig 2001]. In this new frontier, one or more of the classical assumptions are not taken into account. The advent of powerful computers has triggered the development of individual-based models, such as CA, for the study of non-equilibrium systems. Non-equilibrium multi-particle systems provide analogies with the biological multi-cellular ones. For instance, the motion of charged particles under an electrical field can be compared with haptotactical cell motion (see Chapter 5). The analogies of such physical and biological systems are not complete in each detail but they are instructive. In particular, on the level of mathematical abstraction non-equilibrium multi-particle and biological multi-cellular systems are very similar. Therefore, we believe that non-equilibrium statistical mechanics provides a promising framework that allows for the deeper investigation of multi-cellular systems dynamics, such as tumors.

9.3 Discussion and outlook

The research conducted in this thesis can be distinguished in two principle venues: (i) novel analytical techniques for LGCA and (ii) innovative modeling of tumor invasion. In the following, we critically discuss the results of the thesis and we propose possible extensions or improvements.

9.3. Discussion and outlook

9.3.1 Mathematical analysis of LGCA

In the first part of the thesis, we have dealt with analysis of LGCA mainly by means of mean-field analysis. Mean-field analysis allows for an average description of the single node behavior, which corresponds to the LBE. The LBE is the discrete analogous of the Boltzmann equation in classical kinetic theory. From this average node description, we have extrapolated the macroscopic behavior of the system. The mathematical tools that allow for a coarse-grained picture of the system are called scaling methods. In particular, here we have used and discussed three different scaling methods, the Chapman-Enskog, the CML and the Fourier space method.

An important theoretical issue, that has been extensively discussed in this thesis, is the study of a propagating wavefront. In particular, we have been interested in providing analytical estimators of the front speed. For the calculation of the invasion front speed, it has been observed that spatial fluctuations play an important role. As a consequence, the macroscopic description obtained under the mean-field assumption fails at the front. In this thesis, we have proposed the use of a cut-off at the level of LBE. The cut-off mean-field description allows for a better approximation of the dynamics at the front. However, this phenomenological treatment of the front problem suffers from some limitations.

A central point of our mathematical analysis is the mean-field assumption. This simplification of the stochastic dynamics enables us to analyze macroscopically the systems. At the same time, the mean-field assumption neglects all the spatial correlations and we loose a great deal of information concerning the actual microscopic processes. An example that the mean-field approximation (without the cut-off) fails is the exact prediction of the front speed. A straight forward improvement of our mathematical analysis would be the extension of the mean-field approximation by the inclusion of local fluctuations. This will introduce a more sophisticated macroscopic description that would be able to describe more efficiently the system. In this way, one could significantly improve the description of the front dynamics and allow for an exact calculation of the front speed. However, the explicit calculation of the front fluctuations and the exact prediction of the front speed is still a mathematical challenge for discrete, stochastic, spatially distributed processes, such as LGCA.

9.3.2 Tumor invasion modeling

The second venue of research, that has been developed in this thesis, is the modeling of tumor invasion. Tumor invasion is recognized as complex phenomenon. The philosophy of our approach is based on the analysis (from the Greek word $ανά + λύειν$ which means to break down into little pieces) of tumor invasion into basic biological processes. Each process comprises a subproblem - here we call it module. The goal is to develop simple models of the different tumor invasion processes that are mathematically tractable and analyzable. The synthesis of these modules, accompanied with the insight gained from the mathematical analysis of the corresponding models, may provide a deeper understanding of the full problem, i.e. the tumor invasion.

Chapter 9. Discussion

Here, we have reduced the problem to the interplay between migration and proliferation of tumor cells. Our study has shown that the interplay of these two processes is capable of reproducing basic features of tumor invasion. An interesting perspective is the in-depth investigation of the migration/proliferation dichotomy. As we have seen the "Go or Grow" phenomenon may have significant implications for the emergence of tumor invasion. Assuming that the tumor cell population consists of two interacting subpopulations, the resting and the moving cells, we can devise new ideas of modeling and analysis. In particular, a system of moving and resting cells is an extreme Turing system with a population of finite diffusion coefficient and another with zero motility. In such a case, the Turing instability could be expressed in the front the system leading to pattern formation along the boundary. Therefore, the "Go or Grow" mechanism could potentially explain the invasive tumor morphology, i.e. the "fingering" spatial patterns.

Moreover, it is known that populations with different motility rates could induce sub- or super-diffusive phenomena [Peruani 2007]. Since the "Go or Grow" mechanism is based on two-different populations with different motility properties, we can deduce that the tumor invasion exhibits a rich repertoire of behavior. It is of interest to know when and where invasive tumors exhibit anomalous diffusion. Preliminary results show that including the "Go or Grow" mechanism in mathematical models, we can reproduce the behavior of biological experiments concerning tumor cell motility [Chauviere 2009] or growing tumor cell cultures [Tektonidis 2009].

The impact of the surrounding environment on tumor invasion has been also investigated in this thesis, with special focus on the migration of tumor cells. However, the influence of a tumor's environment on tumor invasion can be observed in a multitude of phenomena. An aspect that has not been considered in this thesis is the degradation of the ECM mediated by the tumor cell production of metalloproteinases. This process leads to ECM remodeling establishing a dynamic relationship between tumor cells and ECM. A first modeling approach has been conducted by [Hillen 2006]. It is an open problem which are the rules that govern the tumor-ECM relationship. Of particular interest is the impact of this relationship on tumor invasion dynamics.

To summarize, we hope that this thesis contributes in the understanding of tumor invasion dynamics. Our aim was not only to shed light on aspects of tumor invasion but also to identify appropriate mathematical tools for the analysis of tumor invasion behavior. Additionally, we hope that we could motivate new research ideas that may help in the profound comprehension of the mechanisms of tumor growth and the design of novel therapeutic strategies.

APPENDIX A

Calculation of random walk LBE

Here we compute analytically the rhs of equation (2.18) in section 2.3.1. We first evaluate the number of elements in the equivalence class Z:

$$\begin{aligned}
Z &= \sum_{\boldsymbol{\eta}^O(r,k)} \delta\big(n(\mathbf{r},k), n^O(\mathbf{r},k)\big) = \sum_{\boldsymbol{\eta}^O(r,k)} \delta\big(\sum_j \eta_j(\mathbf{r},k), \sum_j \eta_j^O(\mathbf{r},k)\big) = \\
&= \sum_{\boldsymbol{\eta}^O} \frac{1}{2\pi} \int_0^{2\pi} dx \exp\{\imath \sum_{j=1}^{\tilde{b}} (\eta_j - \eta_j^O)x\} = \frac{1}{2\pi} \int_0^{2\pi} dx \{\imath \sum_{j=1}^{\tilde{b}} (\eta_j - \eta_j^O)x\} = \\
&= \frac{1}{2\pi} \int_0^{2\pi} dx \prod_j [e^{\imath \eta_j x} \sum_{\eta_j^O=0}^{1} e^{\imath \eta_j^O x}] = \frac{1}{2\pi} \int_0^{2\pi} dx \exp\{\imath \sum_j \eta_j x\}(1+e^{-\imath x})^{\tilde{b}} \\
&= \sum_{l=0}^{\tilde{b}} \frac{1}{2\pi} \int_0^{2\pi} dx \exp\{\imath(\sum_j \eta_j - l)x\} = \sum_{l=0}^{\tilde{b}} \binom{\tilde{b}}{l} \delta(l, \sum_j \eta_j) = \binom{\tilde{b}}{\sum_j \eta_j}.
\end{aligned}$$

Thus, the number of configurations with a given number of cells is

$$Z = \binom{\tilde{b}}{\sum_i \eta_i(\mathbf{r},k)} = \binom{\tilde{b}}{n(\mathbf{r},k)}, \tag{A.1}$$

i.e. the number of different ways of placing $n(\mathbf{r},k)$ cells in the $\tilde{b}$ channels. We now evaluate the numerator of the LBE, which is the number of configurations with channel i occupied

$$\begin{aligned}
\sum_{\boldsymbol{\eta}^O} \eta_i^O \delta(\sum_j \eta_j, \sum_j \eta_j^O) &= \frac{1}{2\pi} \int_0^{2\pi} dx \exp\{\imath \sum_{j=1}^{\tilde{b}} \eta_j x\}(1+e^{-\imath x})^{\tilde{b}} \prod_{j\neq i} \sum_{\eta_i^O=0}^{1} \eta_i^O e^{-\imath \eta_i^O x} \\
&= \frac{1}{2\pi} \int_0^{2\pi} dx \exp\{\imath \sum_{j=1}^{\tilde{b}} \eta_j x\}(1+e^{-\imath x})^{\tilde{b}} e^{-\imath x} = \\
&= \binom{\tilde{b}}{\sum_j \eta_j - 1}. \tag{A.2}
\end{aligned}$$

To obtain the above results we have approximated the $\delta(\cdot)$ function by the Cauchy distribution, and the following identity:

$$(1+e^{-\imath x})^k = \sum_{l=0}^{k} \binom{k}{l} e^{-\imath x}. \tag{A.3}$$

Combining the above derived eqs. (A.1) & (A.2), we get:

$$\sum_{\eta^O} \eta_i^O A_{\eta \to \eta^O} = \frac{\sum_{\eta^O} \eta_i^O \delta(n, n^O)}{Z} = \frac{1}{\tilde{b}} \sum_{i=1}^{\tilde{b}} \eta_i(\mathbf{r}, k), \quad (A.4)$$

which just states that after the application of the reorientation operator, the average occupation of a channel is proportional to the number of cells at this node.

APPENDIX B

Calculation of equilibrium distributions

B.1 Equilibrium distribution of Model I

In this appendix, we calculate in detail the equilibrium distribution for model I. For the zero-field case, we know that the equilibrium distribution is $f_i^{eq} = \rho/b = d$. Thus, we can easily find that $h_0 = ln(\frac{1-d}{d})$. For simplicity of the notation we use f_i instead of f_i^{eq}.

The next step is to expand the equilibrium distribution around $\mathbf{E} = \mathbf{0}$ and we obtain:

$$f_i = f_i(\mathbf{E}=0) + (\nabla_\mathbf{E}) f_i \mathbf{E} + \frac{1}{2}\mathbf{E}^T(\nabla_\mathbf{E}^2) f_i \mathbf{E}. \tag{B.1}$$

In the following, we present the detailed calculations. The chain rule gives:

$$\frac{\partial f_i}{\partial e_\alpha} = \frac{\partial f_i}{\partial x} \frac{\partial x}{\partial e_\alpha}. \tag{B.2}$$

Then using eqs. (5.9) and (5.10) :

$$\frac{\partial f_i}{\partial x} = -\frac{e^x}{(1+e^x)^2} \to d(d-1) \tag{B.3}$$

$$\frac{\partial x}{\partial e_\alpha} = \frac{\partial}{\partial e_\alpha}(h_0 + h_1 c_i \mathbf{E} + h_2 \mathbf{E}^2) = h_1 c_{i\alpha} + 2h_2 e_\alpha. \tag{B.4}$$

For $\mathbf{E} = \mathbf{0}$ one obtains:

$$\frac{\partial f_i}{\partial e_\alpha} = d(d-1) h_1 c_{i\alpha}, \tag{B.5}$$

where $\alpha = 1, 2$. Then, we calculate the second order derivatives:

$$\frac{\partial^2 f_i}{\partial e_\alpha^2} = \frac{\partial}{\partial e_\alpha}(\frac{\partial f_i}{\partial x}\frac{\partial x}{\partial e_\alpha}) = \frac{\partial^2 f_i}{\partial x \partial e_\alpha}\frac{\partial x}{\partial e_\alpha} + \frac{\partial f_i}{\partial x}\frac{\partial^2 x}{\partial e_\alpha^2} = \frac{\partial^2 f_i}{\partial x^2}(\frac{\partial x}{\partial e_\alpha})^2 + \frac{\partial f_i}{\partial x}\frac{\partial^2 x}{\partial e_\alpha^2}. \tag{B.6}$$

Especially:

$$\frac{\partial^2 f_i}{\partial x^2} = \frac{e^x(e^x-1)}{(1+e^x)^3} = d(d-1)(2d-1) \tag{B.7}$$

$$\frac{\partial^2 x}{\partial e_\alpha^2} = 2h_2. \tag{B.8}$$

Thus, relation (B.6) reads:

$$\frac{\partial^2 f_i}{\partial e_\alpha^2} = d(d-1)(2d-1)h_1^2 c_{i\alpha} + d(d-1)2h_2. \qquad (B.9)$$

For the case $\alpha \neq \beta$ ($\alpha, \beta = 1, 2$), we have:

$$\frac{\partial^2 f_i}{\partial e_\alpha \partial e_\beta} = \frac{\partial}{\partial e_\beta}(\frac{\partial f_i}{\partial x}\frac{\partial x}{\partial e_\alpha}) = \frac{\partial^2 f_i}{\partial x \partial e_\beta}\frac{\partial x}{\partial e_\alpha} + \frac{\partial f_i}{\partial x}\frac{\partial^2 x}{\partial e_\alpha \partial e_\beta} = \frac{\partial^2 f_i}{\partial x^2}\frac{\partial x}{\partial e_\alpha}\frac{\partial x}{\partial e_\beta} + \frac{\partial f_i}{\partial x}\frac{\partial^2 x}{\partial e_\alpha \partial e_\beta}. \qquad (B.10)$$

We can easily derive:

$$\frac{\partial^2 x}{\partial e_\alpha \partial e_\beta} = 0. \qquad (B.11)$$

Thus, equation (B.10) becomes:

$$\frac{\partial^2 f_i}{\partial e_\alpha \partial e_\beta} = d(d-1)(2d-1)h_1^2 c_{i\alpha} c_{i\beta}. \qquad (B.12)$$

Finally, the equilibrium distribution reads (approximation up ot second order in **E**:

$$f_i = d + d(d-1)h_1 \mathbf{c}_i \mathbf{E} + \frac{1}{2}d(d-1)(2d-1)h_1^2 \sum_\alpha c_{i\alpha}^2 e_\alpha^2 + d(d-1)h_2 \mathbf{E}^2. \qquad (B.13)$$

In the last relation, we have to determine the free parameters h_1, h_2. Using the mass conservation law, we can find a relation between h_1 and h_2:

$$\rho = \sum_{i=1}^b f_i = \underbrace{\sum_i d}_{\rho} + d(d-1)h_1 \underbrace{\sum_i \mathbf{c}_i \mathbf{E}}_{0}$$

$$+ \frac{1}{2}d(d-1)(2d-1)h_1^2 \underbrace{\sum_i \sum_\alpha c_{i\alpha}^2 e_\alpha^2}_{\frac{b}{2}\mathbf{E}^2}$$

$$+ d(d-1)h_2 \sum_i h_2 \mathbf{E}^2. \qquad (B.14)$$

For any choice of the lattice, we find:

$$h_2 = \frac{1-2d}{4}h_1^2. \qquad (B.15)$$

Finally, the equilibrium distribution can be explicitly calculated for small driving fields:

$$f_i = d + d(d-1)h_1 \mathbf{c}_i \mathbf{E} + \frac{1}{2}d(d-1)(2d-1)h_1^2 Q_{\alpha\beta} e_\alpha e_\beta, \qquad (B.16)$$

where $Q_{\alpha\beta} = c_{i\alpha} c_{i\beta} - \frac{1}{2}\delta_{\alpha\beta}$ is a second order tensor.

We now calculate the mean flux, in order to obtain the linear response relation:

$$\langle \mathbf{J}(\eta^C) \rangle = \sum_i c_{i\alpha} f_i = \frac{b}{2} d(d-1) h_1 \mathbf{E}. \qquad (B.17)$$

Thus, the susceptibility reads:

$$\chi = \frac{1}{2} b d(d-1) h_1 = -\frac{1}{2} b g_{eq} h_1. \qquad (B.18)$$

B.2 Equilibrium distribution of Model II

In this appendix, we present details of the calculation of the equilibrium distribution for model II. To simplify the calculations, we restrict ourselves to the case of the square lattice ($b=4$).

Using the mass conservation law, allows to calculate the relation between h_1, h_2.

$$\begin{aligned}
\rho = \sum_{i=1}^{b} f_i &= \underbrace{\sum_i d}_{\rho} + d(d-1) h_1 \sum_i |\mathbf{c}_i| \mathbf{E} \\
&+ \frac{1}{2} d(d-1)(2d-1) h_1^2 \underbrace{\sum_i \sum_\alpha c_{i\alpha}^2 e_\alpha^2}_{\frac{b}{2} E^2} \\
&+ d(d-1)(2d-1) h_1^2 \underbrace{\sum_i |c_{i\alpha} c_{i\beta} e_\alpha| e_\beta}_{\frac{b}{2} \delta_{\alpha\beta} e_\alpha e_\beta} \\
&+ d(d-1) h_2 \sum_i \mathbf{E}^2. \qquad (B.19)
\end{aligned}$$

Finally, the previous equation becomes:

$$2d(d-1) h_1 \sum_\alpha e_\alpha + d(d-1)(2d-1) h_1^2 \mathbf{E}^2 + 4d(d-1) h_2 \mathbf{E}^2 = 0, \qquad (B.20)$$

and we find:

$$h_2 = \frac{1-2d}{4} h_1^2 - \frac{1}{2} \frac{e_1 + e_2}{e_1^2 + e_2^2} h_1. \qquad (B.21)$$

B.3 Analytical parameter estimation in Model II

In this appendix, we estimate the free parameter h_1 for model II.

The field $\mathbf{E}$ induces a spatially homogeneous deviation from the field-free equilibrium state $f_i(\mathbf{r}|\mathbf{E}=0) = f_{eq}$ of the form:

$$f_i(\mathbf{r}|\mathbf{E}) = f_{eq} + \delta f_i(\mathbf{E}). \qquad (B.22)$$

150 Appendix B. Calculation of equilibrium distributions

We denote the transition probability as $\mathbb{P}(\boldsymbol{\eta} \to \boldsymbol{\eta}^C) = A_{\eta\eta^C}$. The average flux is given by:

$$\langle \mathbf{J}(\eta^C) \rangle = \sum_{i=1}^{b} \mathbf{c}_i \delta f_i(\mathbf{E}). \tag{B.23}$$

For small $\mathbf{E}$ we expand eq. (5.4) as:

$$A_{\eta\eta^C}(\mathbf{E}) \simeq A_{\eta\eta^C}(0)\{1 + [|\mathbf{J}(\eta^C)| - |\overline{\mathbf{J}(\eta^C)}|]\mathbf{E}\}, \tag{B.24}$$

where we have defined the expectation value of $\mathbf{J}(\eta^C)$) averaged over all possible outcomes η^C of a collision taking place in a field-free situation:

$$|\overline{\mathbf{J}(\eta^C)}| = \sum_{\eta^C} |\mathbf{J}(\eta^C)| A_{\eta\eta^C}(0). \tag{B.25}$$

In the mean-field approximation the deviations $\delta f(\mathbf{E})$ are implicitly defined as stationary solutions of the non-linear Boltzmann equation for a given $\mathbf{E}$, i.e.

$$\Omega_i^{10}[f_{eq} + \delta f_i(\mathbf{E})] = 0. \tag{B.26}$$

Here the nonlinear Boltzmann operator is defined by:

$$\Omega_i^{10}(\mathbf{r},t) = \langle \eta_i^C(\mathbf{r},t) - \eta_i(\mathbf{r},t) \rangle_{MF} = \sum_{\eta^C} \sum_{\eta} [\eta_i^C(\mathbf{r},t) - \eta_i(\mathbf{r},t)] A_{\eta\eta^C}(\mathbf{E}) F(\eta,\mathbf{r},t), \tag{B.27}$$

where the factorized single node distribution is defined as:

$$F(\eta,\mathbf{r},t) = \prod_i [f_i(\mathbf{r},t)]^{\eta_i}[1 - f_i(\mathbf{r},t)]^{1-\eta_i}. \tag{B.28}$$

Linearizing around the equilibrium distribution yields:

$$\Omega_i^{10}[f_{eq} + \delta f_i(\mathbf{E})] = \Omega_i^{10}(f_i) + \sum_j \Omega_{ij}^{11}(f_{eq})\delta f_j(\mathbf{E}), \tag{B.29}$$

where $\Omega_{ij}^{11} = \frac{\partial \Omega_i^{10}}{\partial f_j}$. Moreover:

$$\Omega_i^{10}(f_{eq}) = \sum_{\eta^C,\eta}(\eta_i^C - \eta_i)\{1 + [|\mathbf{J}(\eta^C)| - |\overline{\mathbf{J}(\eta^C)}|]\mathbf{E}\}A_{\eta\eta^C}(0)F(\eta). \tag{B.30}$$

Using the relations $\sum(\eta_i^C - \eta_i)A_{\eta\eta^C}(0)F(\eta) = 0$ and $\sum(\eta_i^C - \eta_i)|\overline{\mathbf{J}(\eta^C)}|A_{\eta\eta^C}(0)F(\eta) = 0$, we obtain:

$$\Omega_i^{10}(f_{eq}) = \langle (\eta_i^C - \eta_i)|\mathbf{J}(\eta^C)|\rangle \mathbf{E}. \tag{B.31}$$

Around $\mathbf{E} = \mathbf{0}$ one obtains:

$$\sum_j \Omega_{ij}^{11}(f_{eq})\delta f_j(\mathbf{E}) + \langle (\eta_i^C - \eta_i)|\mathbf{J}(\eta^C)|\rangle_{MF}\mathbf{E} = 0. \tag{B.32}$$

B.3. Analytical parameter estimation in Model II

Solving the above equation involves the inversion of the symmetric matrix $\Omega^{11}_{ij} = 1/b - \delta_{ij}$. It can be proven that the linearized Boltzmann operator looks like:

$$\Omega^{11}_{ij} = \langle(\delta\eta^C_i - \delta\eta_i)\frac{\delta\eta_j}{g_{eq}}\rangle = \frac{1}{g_{eq}}(\langle\delta\eta^C_i, \delta\eta_j\rangle - \langle\delta\eta_i, \delta\eta_j\rangle), \quad (B.33)$$

where $\delta\eta_i = \eta_i - f_{eq}$ and the single particle fluctuation $g_{eq} = f_{eq}(1 - f_{eq})$. For the second term of the last part of eq. (B.33), we have $\langle\delta\eta_i, \delta\eta_j\rangle = \delta_{ij}g_{eq}$. To evaluate the first term, we note that the outcome of the collision rule only depends on $\eta(\mathbf{r})$ through $\rho(\mathbf{r})$, so that the first quantity does not depend on the i and j and

$$\langle\delta\eta^C_i, \delta\eta_j\rangle = \frac{1}{b^2}\langle[\delta\rho(\mathbf{r})]^2\rangle = \frac{1}{b}g_{eq}, \quad (B.34)$$

where we have used $\rho(\eta) = \rho(\eta^C)$. Thus eq. (B.33) takes the value $(1/b - \delta_{ij})$.

Returning to the calculation of the generalized inverse of Ω^{11}, we observe that its null space is spanned by the vector $(\overbrace{1,..,1}^{b})$, which corresponds to the conservation of particles

$$\sum_i \delta f_i(\mathbf{E}) = 0. \quad (B.35)$$

The relation satisfies the solvability condition of the Fredholm alternative for eq. (B.32), which enables us to invert the matrix within the orthogonal complement of the null space. With some linear algebra, we can prove that the generalized inverse $[\Omega^{11}]^{-1}$ has the same eigenvectors but inverse eigenvalues as the original matrix Ω^{11}. In particular, it can be verified that since $\mathbf{c}_{\alpha i}$, $\alpha = 1, 2$ (where 1, 2 stands for x- and y-axis, respectively) are eigenvectors of Ω^{11} with eigenvalue -1, we have

$$\sum_j [\Omega^{11}_{ij}]^{-1} c_{aj} = -c_{ai}. \quad (B.36)$$

Now we can calculate the flux of particles for one direction:

$$\langle J_{a^+}(\eta^C)\rangle = -\sum_j c_{aj}[\Omega^{11}_{ij}]^{-1}\langle(\eta^C_i - \eta_i)|\mathbf{J}_\beta(\eta^C)|\rangle e_a$$

$$= c_{ai}\langle(\eta^C_i - \eta_i)|\mathbf{J}_\beta(\eta^C)|\rangle e_a. \quad (B.37)$$

Calculating in detail the last relation:

$$c_{ai}\langle(\eta^C_i - \eta_i)|\mathbf{J}_\beta(\eta^C)|\rangle = c_{ai}\sum_j |c_{\beta j}|\langle(\delta\eta^C_i - \delta\eta_i)\delta\eta^C_j\rangle \quad (B.38)$$

$$= \frac{1}{2}g_{eq}c_{ai}. \quad (B.39)$$

The observable quantity that we want to calculate for the second rule is:

$$|\langle\mathbf{J}_{x^+}(\eta^C)\rangle - \langle\mathbf{J}_{y^+}(\eta^C)\rangle| = \frac{1}{2}g_{eq}|e_1 - e_2|, \quad (B.40)$$

since $c_{11} = c_{22} = 1$.

APPENDIX C

Details of tumor growth model in Chapter 6

In this Appendix, we present the details of the microdynamical eqs (6.4) and (6.5). In the following for simplicity reasons and without any loss of generality, we drop the spatial and the temporal arguments of the functions. The Heaviside functions $\Theta(\theta_M - n_C)$ and $\Theta(n_C - \theta_N)$ can be alternatively written in terms of random variables:

$$\Theta(\theta_M - n_C) = \sum_{l=1}^{\theta_M} \delta(n_C = l) = \begin{cases} 1, & \text{if } n_C \leq \theta_M \\ 0, & \text{else} \end{cases}$$

$$\Theta(n_C - \theta_N) = \sum_{l=\theta_N}^{\tilde{b}} \delta(n_C = l) = \begin{cases} 1, & \text{if } n_C \geq \theta_N \\ 0, & \text{else} \end{cases}$$

where the $\delta(n_C)$ functions represent the possible node configurations that account for n_C number of cells, defined in the general form:

$$\delta(n = N) = \sum_{l=1}^{\binom{\tilde{b}}{N}} \prod_{i \in M_l^n} \eta_i(r,k) \prod_{j \in M/M_l^n} (1 - \eta_j(r,k)) = \begin{cases} 1, & \text{if } n = N \\ 0, & \text{else} \end{cases} \quad (C.1)$$

where $N \in \{0, ..., \tilde{b}\}$, the index set $M = 1, ..., \tilde{b}$ and M_l^n denoting the nth subset of M with l elements.

Now let us evaluate the expected collision operators $\tilde{C}_{\sigma,i}$ from eqs. (6.9) and (6.10). We assume that $\theta_M = 4$ and $\theta_N = 6$ and that the system is in the steady state $(\bar{f}_C, \bar{f}_N)$. Moreover, we fix the node capacity as $\tilde{b} = 4$. Therefore, eqs. (6.9) and (6.10) yield:

$$\tilde{C}_{C,i} = \frac{1}{8}\left[r_M\left(\bar{f}_C(1-\bar{f}_C)^7 + 28\bar{f}_C^2(1-\bar{f}_C)^6 + 56\bar{f}_C^3(1-\bar{f}_C)^5 + 70\bar{f}_C^4(1-\bar{f}_C)^4\right)\right.$$
$$\left. - r_N\left(28\bar{f}_C^6(1-\bar{f}_C)^2 + 8\bar{f}_C^7(1-\bar{f}_C)^1 + \bar{f}_C^8\right)\right], \quad (C.2)$$

$$\tilde{C}_{N,i} = \frac{1}{8}\left[r_N(1-\bar{f}_N)\left(28\bar{f}_C^6(1-\bar{f}_C)^2 + 8\bar{f}_C^7(1-\bar{f}_C)^1 + \bar{f}_C^8\right)\right]. \quad (C.3)$$

Setting $\tilde{C}_{\sigma,i} = 0$, we can calculate the exact values of the steady states $(\bar{f}_C, \bar{f}_N)$ in (6.11).

Bibliography

[Alarcón 2003] T. Alarcón, H. M. Byrne and P. K. Maini. *A cellular automaton model for tumour growth in inhomogeneous environment.* J. Theor. Biol., vol. 225, no. 2, pages 257–274, 2003. 5

[Alexander 1992] F. J. Alexander, I. Edrei, P. L. Garrido and J. L. Lebowitz. *Phase transitions in a probabilistic cellular automaton: growth kinetics and critical properties.* J. Statist. Phys., vol. 68, no. 3/4, page 497514, 1992. 67

[Alt 1980] W. Alt. *Biased random walk models for chemotaxis and related diffusion approximations.* J. Math. Biol., vol. 9, page 147177, 1980. 22

[Anderson 2005] A. R. A. Anderson. *A hybrid model of solid tumour invasion: the importance of cell adhesion.* Math. Med. Biol., vol. 22, pages 163–186, 2005. 118

[Anderson 2006] A. R. Anderson, A. M. Weaver, P. T. Cummings and V. Quaranta. *Tumor morphology and phenotypic evolution driven by selective pressure from the microenvironment.* Cell, vol. 127, no. 5, pages 905–15, 2006. 81, 118

[Athale 2006] C. Athale, Y. Mansury and T. Deisboeck. *Simulating the impact of a molecular decision-process on cellular phenotpe and multicellular patterns in brain tumors.* J. Theor. Biol., vol. 239, pages 516–527, 2006. 103

[Banks 1994] R. B. Banks. *Growth and diffusion phenomena: Mathematical frameworks and applications.* Springer, New York, 1994. 41

[Basanta 2008a] D. Basanta, H. Hatzikirou and A. Deutsch. *he emergence of invasiveness in tumours: a game theoretic approach.* Eur. Phys. J. B, vol. 63, page 393397, 2008. 5, 81, 118

[Basanta 2008b] D. Basanta, M. Simon, H. Hatzikirou and A. Deutsch. *An evolutionary game theory perspective elucidates the role of glycolysis in tumour invasion.* Cell Prolif., vol. 41, page 980987, 2008. 118

[Benguria 2004] R. D. Benguria, M. C. Depassier and V. Mendez. *Propagation of fronts of a reaction-convection-diffusion equation.* Phys. Rev. E, vol. 69, page 031106, 2004. 97

[Bentley 1999] P. J. Bentley. *Evolutionary design by computers.* Morgan Kaufmann Publishers, Inc., 1999. 123

[Blossey 2006] R. Blossey. *Computational biology: A statistical mechanics perspective.* Chapman Hall/Crc Mathematical and Computational Biology Series, 2006. 141

Bibliography

[Boerlijst 2006] M. Boerlijst. *Book Review of "Cellular automaton modeling of biological pattern formation"*. Math. Biosc., vol. 200, pages 118–123, 2006. 43

[Boon 1996] J. P. Boon, D. Dab, R. Kapral and A. Lawniczak. *Lattice Gas Automata for Reactive Systems*. Phys. Rpts., vol. 273, pages 55–148, 1996. 93

[Bray 1992] D. Bray. Cell movements. Garland Publishing, New York, 1992. 21

[Breuer 1994] H. P. Breuer, W. Huber and F. Petruccione. *Fluctuation effects on wave propagation in a reaction-diffusion process*. Phys. D, vol. 73, page 259, 1994. 94

[Bru 2003] A. Bru, S. Albertos, J. L. Subiza, J. Lopez Garcia-Asenjo and I. Bru. *The universal dynamics of tumor growth*. Bioph. J., vol. 85, pages 2948–2961, 2003. 6, 99, 100

[Brunet 1997] I. Brunet and B. Derrida. *Shift in the velocity of a front due to a cutoff*. Phys. Rev. E, vol. 56, no. 3, pages 2597–2604, 1997. 94, 97

[Brunet 2001] I. Brunet and B. Derrida. *Effect of microscopic noise in front propagation*. J. Stat. Phys., vol. 103, no. 1/2, pages 269–282, 2001. 97

[Bussemaker 1996] H. Bussemaker. *Analysis of a pattern forming lattice gas automaton: mean field theory and beyond*. Phys. Rev. E, vol. 53, no. 4, page 16441661, 1996. 77, 80

[Byrne 2003] H. Byrne and L. Preziosi. *Modelling solid tumour growth using the theory of mixtures*. Math. Medic. and Biol., vol. 20, no. 4, pages 341–366, 2003. 64

[Carter 1965] S. B. Carter. *Principles of cell motility: the direction of cell movement and cancer invasion*. Nature, vol. 208, no. 5016, pages 1183–1187, 1965. 63

[Casey 1934] A. E. Casey. *The experimental alterations of malignancy with a homologous mammalian tumor material*. Am. J. Canc., vol. 24, pages 760–775, 1934. 42

[Chauviere 2007] A. Chauviere, T. Hillen and L. Preziosi. *Modeling the motion of a cell population in the extracellular matrix*. Discr. Cont. Dyn. Syst., vol. Suppl. vol, pages 250–259, 2007. 64

[Chauviere 2009] A. Chauviere, H. Hatzikirou and A. Deutsch. *Anomalous diffusion of cell migration: the "Go or Rest" model*. 2009. in preparation. 144

[Chopard 1998] B. Chopard and M. Droz. Cellular automata modeling of physical systems. Cambridge University Press, Cambridge, 1998. 25, 31, 32, 49

[Clark 1998] M. C. Clark, L. O. Hall, D. B. Goldgof, R. Velthuizen, F. R. Murtagh and M. S. Silbiger. *Automatic tumor segmentation using knowledge-based techniques*. IEEE transactions on medical imaging, vol. 17, no. 2, pages 187–201, 1998. 122, 126

[Cohen 2005] E. Cohen, D. Kessler and H. Levine. *Fluctuation-regularized front propagation dynamics in reaction-diffusion systems*. Phys. Rev. Lett., vol. 94, page 158302, 2005. 94

[Dab 1991] D. Dab, J. Boon, and Y. X. Li. *Lattice-gas for coupled reaction-diffusion equations*. Phys. Rev. Lett., vol. 66, pages 2535–2538, 1991. 84

[Dallon 2001] J. C. Dallon, J. A. Sherratt and P. K. Maini. *Modelling the effects of transforming growth factor on extracellular alignment in dermal wound repair*. Wound Rep. Reg., vol. 9, pages 278–286, 2001. 64

[Daruwalla 2006] J. Daruwalla and C. Christophi. *Hyperbaric oxygen therapy for malignancy: a review*. World J. Surg., vol. 30, no. 12, pages 2112–31, 2006. 118, 119

[Dembo 1989] M. Dembo. *Field theorems of the cytoplasm*. Comments Theor. Biol., vol. 1, page 159177, 1989. 21

[Deutsch 2005] A. Deutsch and S. Dormann. *Cellular automaton modeling of biological pattern formation*. Birkhäuser, 2005. 5, 6, 25, 43, 64, 139

[Dickinson 1993] R. B. Dickinson and R. T. Tranquillo. *A stochastic model for cell random motility and haptotaxis based on adhesion receptor fuctuations*. J. Math. Biol., vol. 31, pages 563–600, 1993. 21, 22, 64, 70

[Dickinson 1995] R. B. Dickinson and R. T. Tranquillo. *Transport equations and cell movement indices based on single cell properties*. SIAM J. Appl. Math., vol. 55, no. 5, pages 1419–54, 1995. 64

[Dolak 2005] Y. Dolak and C. Schmeiser. *Kinetic models for chemotaxis: Hydrodynamic limits and spatio-temporal mechanics*. J. Math. Biol., vol. 51, pages 595–615, 2005. 64

[Doucet 1992] P. Doucet and P.B. Sloep. Mathematical modeling in the life sciences. Ellis Horwood, New York, 1992. 41

[Dunn 1987] G. A. Dunn and A. F. Brown. *A unified approach to analyzing cell motility*. J. Cell Sci. Suppl., vol. 8, page 81102, 1987. 22

[Eckerich 2007] C. Eckerich, S. Zapf, R. Fillbrandt, S. Loges, M. Westphal and K. Lamszus. *Hypoxia can induce c-Met expression in glioma cells and enhance SF/HGF-induced cell migration*. Int J Cancer., vol. 15, no. 121, pages 276–83, 2007. 119

[Einstein 1905] A. Einstein. *Ueber die von der molekulartheoretischen Theorie der Waerme geforderte Bewegung von in ruhenden Fluessigkeiten suspendierten Teilchen.* Ann. Phys., vol. 17, page 549, 1905. 23

[Elton 1995] H. Elton, C. Levermore and G. Rodrigue. *Convergence of convective-diffusive lattice Boltzmann methods.* SIAM J. Numer. Anal., vol. 32, no. 5, pages 1327–1354, 1995. 32

[Fedotov 2007] S. Fedotov and A. Iomin. *Migration and proliferation dichotomy in tumor-cell invasion.* Phys. Rev. Let., vol. 98, pages 118101–4, 2007. 103

[Folkman 1973] J. Folkman and M. Hochberg. *Self-Regulation of growth in three dimensions.* J. Exp. Med., vol. 138, pages 745–753, 1973. 87, 99, 177

[Ford 1991] R. M. Ford and D. A. Lauffenburger. *Measurement of bacterial random motility and chemotaxis coefficients: II. Application of single-cell-based mathematical model.* Biotech. Bioeng., vol. 37, page 661672, 1991. 22

[Frieboes 2007] H. B. Frieboes, S. Wise, X. Zheng, P. Macklin, E. Bearer and V. Cristini. *Computer simulation of glioma growth and morphology.* NeuroImage, vol. 37, pages 59–70, 2007. 5

[Friedl 2000] P. Friedl and E.B. Broecker. *The biology of cell locomotion within a three dimensional extracellular matrix.* Cell Motility Life Sci., vol. 57, pages 41–64, 2000. 61, 62

[Friedl 2003] P. Friedl and K. Wolf. *Tumour-cell invasion and migration: diversity and escape mechanisms.* Nature Rev., vol. 3, pages 362–374, 2003. 63

[Friedl 2004] P. Friedl. *Prespecification and plasticity: shifting mechanisms of cell migration.* Curr. Opin. Cell. Biol., vol. 16, no. 1, pages 14–23, 2004. 2, 3, 62, 63, 80

[Frisch 1987] U. Frisch, D. d'Humieres, B. Hasslacher, P. Lallemand, Y. Pomeau and J. P. Rivet. *Lattice gas hydrodynamics in two and three dimensions.* Compl. Syst., vol. 1, pages 649–707, 1987. 11, 76, 127

[Fujiwara 2007] S. Fujiwara, K. Nakagawa, H. Harada, S. Nagato, K. Furukawa, M. Teraoka, T. Seno, K. Oka, S. Iwata and T. Ohnishi. *Silencing hypoxia-inducible factor-1α inhibits cell migration and invasion under hypoxic environment in malignant gliomas.* Int. J. Oncol., vol. 30, no. 4, pages 793–802, 2007. 119

[Gardiner 1990] C. W. Gardiner. *Handbook of stochastic methods.* Springer, Berlin, 1990. 43

[Gatenby 2004] Gatenby and R.J. R.A. Gillies. *Why do cancers have high aerobic glycolysis?* Nat. Rev. Canc., vol. 4, 2004. 102

[Giese 1996a] A. Giese, L. Kluwe, B. Laube and M. E. Berens. *Migration of human glioma cells on myelin.* Neuros., vol. 38, 1996. 102

[Giese 1996b] A. Giese, M. A. Loo, D. Tran, S. W. Haskett and B. M. E. Coons. *Dichotomy of astrocytoma migration and proliferation.* Int. J. Cancer, vol. 67, pages 275–282, 1996. 102

[Giese 2003] A. Giese, R. Bjerkvig, M.E. Berens and M. Westphal. *Cost of Migration: Invasion of Malignant Gliomas and Implications for Treatment.* J. Clin. Onc., vol. 21, no. 8, pages 1624–1636, 2003. 102, 111, 117, 119

[Goldberg-Zimring 2005] D. Goldberg-Zimring, I.-F. Talos, J. G. Bhagwat, S. J. Haker, P.M. Black and K.H. Zou. *Statistical validation of brain tumor shape approximation via spherical harmonics for image-guided neurosurgery.* Academic Radiology Journal, vol. 12, pages 459–466, 2005. 126

[Goldstein 1951] S. Goldstein. *On diffusion by discontinuous movements, and on the telegraph equation.* J. Mech. Appl. Math., vol. 6, 1951. 35

[Grima 2007] R. Grima. *Directed cell migration in the presence of obstacles.* Theor. Biol. Med. Model, vol. 4, no. 2, 2007. doi:10.1186/1742-4682-4-2. 64

[Hanahan 2000] D. Hanahan and R. Weinberg. *The hallmarks of cancer.* Cell, vol. 100, pages 57–70, 2000. 1, 2, 101, 118, 175

[Hardy 1973] J. Hardy and Y. Pomeau und O. de Pazzis. *Time Evolution of a Two-Dimensional Classical Lattice System.* Phys. Rev. Lett., vol. 31, pages 276 – 279, 1973. 11

[Hatzikirou] H. Hatzikirou, B. Basanta, M. Simon, C. Schaller and A. Deutsch. *"Go or Grow": the key to the emergence of invasion in tumor progression?* in preparation. 100

[Hatzikirou 2005] H. Hatzikirou, A. Deutsch, C. Scaller, M. Simon and K. Swanson. *Mathematical modelling of glioblastoma tumour development: a review.* Math. Mod. Meth. Appl. Sc., vol. 15, no. 11, pages 1779–1794, 2005. 4, 64, 105, 121

[Hatzikirou 2008a] H. Hatzikirou, L. Brusch, C. Schaller, M. Simon and A. Deutsch. *Prediction of traveling front behavior in a lattice-gas cellular automaton model for tumor invasion.* Comput. Math. Appl., 2008. in print. 58, 74, 75, 177

[Hatzikirou 2008b] H. Hatzikirou and A. Deutsch. *Cellular automata as microscopic models of cell migration in heterogeneous environments.* Curr. Top. Dev. Biol., vol. 81, pages 401–434, 2008. 55, 100

[Hillen 2006] T. Hillen. *(M5) Mesoscopic and macroscopic models for mesenchymal motion.* J. Math. Biol., vol. 53, page 585616, 2006. 32, 64, 144

[Jähne 2004] B. Jähne. Practical handbook on image processing for scientific and technical applications. CRC Press, 2004. 122

[Jiang 2004] C. Jiang, X. Zhang, W. Huang and C. Meinel. *Segmentation and quantification of brain tumour.* IEEE International Conference on Virtual Environments, Human-Computer Interfaces, and Measurement Systems, Boston, MD, 2004. 126

[Kac 1974] M. Kac. *A stochastic model related to the telegraphers equation.* Rocky Mt. J. Math., vol. 4, 1974. 35

[Kapral 1997] R. Kapral and X. Wu. *Spatiotemporal dynamics of mesoscopic chaotic systems.* Phys. D, vol. 103, pages 314–329, 1997. 46

[Keller 1971] E. F. Keller and L. A. Segel. *Traveling bands of chemotactic bacteria: A theoretical analysis.* J. Theor. Biol., vol. 30, page 235248, 1971. 22, 64

[Kizaka 2003] S. Kizaka, M. Inoue, H. Harada and M. Hiroka. *Tumor hypoxia: A target for selective cancer therapy.* Canc. Sc., vol. 94, no. 12, pages 1021–1028, 2003. 119

[Kot 1992] M. Kot. *Discrete-time travelling waves: ecological examples.* J. of Math. Biol., vol. 30, no. 4, pages 413–436, 1992. 33

[Kubo 1980] R. Kubo. *Brownian Motion and Nonequilibrium Statistical Mechanics.* Science, vol. 233, no. 4761, pages 330 – 334, 1980. 30

[Lamine 2005] K. Lamine, M. Lambin and C. Alauzet. *Effect of starvation on the searching path of the predatory bug deraeocoris lutescens.* BioContr., vol. 50, no. 5, pages 717–727, 2005. 119

[Lauffenburger 1993] D. A. Lauffenburger and J. J. Linderman. Receptors: Models for binding, tracking and signalling. Oxford University Press, New York, 1993. 21

[Lesne 2008] A. Lesne. *Discrete vs continuous controversy in physics.* Math. Struct. Comp. Sc., 2008. in print. 6

[Liggett 1985] T. M. Liggett. Interacting particle systems. Springer-Verlag, Berlin, 1985. 64

[Lundström 1997] C. Lundström. Segmentation of medical image volumes. Master's Thesis in Biomedical Engineering, Linköping University, Sweden, 1997. 126

[Magie 1963] W. F. Magie. A source book in physics. Harvard, 1963. 23

[Mansury 2006] Y. Mansury, M. Diggory and T. Deisboeck. *Evolutionary game theory in an agent-based brain tumor model: Exploring the Genotype-Phenotype link.* J. Theor. Biol., vol. 238, pages 146–156, 2006. 5, 103

[Marchant 2000] B. P. Marchant, J. Norbury and A. J. Perumpanani. *Traveling shock waves arising in a model of malignant invasion.* Journal of Applied Mathematics, vol. 60, pages 263–276, 2000. 4

[McCarthy 1984] J. B. McCarthy and L. T. Furcht. *Laminin and fibronectin promote the haptotactic migration of B16 mouse melanoma cells.* J. Cell Biol., vol. 98, no. 4, pages 1474–80, 1984. 63

[Medlock 2003] J. Medlock and M. Kot. *Spreading disease: integro-differential equations old and new.* Math. Biosc., vol. 184, pages 201–222, 2003. 33

[Merlo 2002] L. Merlo, J. Pepper, B. Reid and C. Maley. *Cancer as an evolutionary and ecological process.* Nat. Rev. Canc., vol. 2, page 924935, 2002. 118

[Mitchell 1996] M. Mitchell. An introduction to genetic algorithms. The MIT Press, 1996. 123

[Mogilner 1996] A. Mogilner and G. Oster. *The physics of lamellipodial protrusion.* Europ. Biophys. J., vol. 25, page 4753, 1996. 21

[Murray 1983] J. D. Murray, G. F. Oster and A. K. Harris. *A mechanical model for mesenchymal morphogenesis.* J. Math. Biol., vol. 17, pages 125–129, 1983. 64

[Murray 2001] J. Murray. Mathematical biology i: An introduction. Springer, 2001. 56, 97

[Neubert 1995] M. G. Neubert, M. Kot and M. A. Lewis. *Dispersal and pattern formation in a discrete-time predator-prey model.* Theoretical Population Biology, vol. 48, no. 1, pages 7–43, 1995. 33

[Newman 2004] T. J. Newman and R. Grima. *Many-body theory of chemotactic cell-cell interactions.* Phys. Rev. E, vol. 70, page 051916, 2004. 64

[Norris 1997] J. R. Norris. Markov chains. Cambridge University Press, 1997. 43

[Nowell 1976] P. C. Nowell. *The clonal evolution of tumor cell populations.* Science, vol. 194, pages 23–28, 1976. 1

[Okubo 2002] A. Okubo and S. A. Levin. Diffusion and ecological problems: modern perspectives. Springer-Verlag, New York, 2002. 64

[Othmer 1988] H. G. Othmer, S. R. Dunbar and W. Alt. *Models of dispersal in biological systems.* J. Math. Biol., vol. 26, pages 263–298, 1988. 35, 64

[Othmer 1997] H. G. Othmer and A. Stevens. *Aggregation, blowup and collapse: The ABCs of taxis in reinforced random walks.* SIAM J. Appl. Math., vol. 57, page 10441081, 1997. 64

[Palecek 1997] S. P. Palecek, J. C. Loftus, M. H. Ginsberg, D. A. Lauffenburger and A. F. Horwitz. *Integrin-ligand binding governs cell-substratum adhesiveness.* Nature, vol. 388, no. 6638, page 210, 1997. 78

[Peruani 2007] F. Peruani and L. Morelli. *Self-propelled particles with fluctuating speed.* Phys. Rev. Lett., vol. 99, page 010602, 2007. 64, 144

[Perumpanani 1996] A. J. Perumpanani, J. A. Sherratt, J. Norbury and H. M. Byrne. *Biological inferences from a mathematical model of malignant invasion.* Invas. Metast., vol. 16, pages 209–221, 1996. 4

[Perumpanani 1999] A. J. Perumpanani, J. A. Sherratt, J. Norbury and H. M. Byrne. *A two parameter family of traveling waves with a singular barrier arising from the modelling of extracellular matrix mediated cellular invasion.* Phys. D, vol. 126, pages 145–159, 1999. 4

[Philibert 1990] J. Philibert. Diffusion et transport de matière dans les solides. (Les Editions de Physique, Les Ulis, France, 1990. 28

[Pratt 2007] W. K. Pratt. Digital image processing. Wiley, 2007. 122

[Preziosi 2003] L. Preziosi. Cancer modelling and simulation. Chapman & Hall/CRC, 2003. 4

[Quaranta 2008] V. Quaranta, K. A. Rejniakb, P. Gerleeb and A. R. A. Anderson. *Invasion emerges from cancer cell adaptation to competitive microenvironments: Quantitative predictions from multiscale mathematical models.* Seminars in Cancer Biology, 2008. doi:10.1016/j.semcancer.2008.03.018. 118

[Rothman 1994] D. H. Rothman and S. Zaleski. *Lattice-gas models of phase separation: interfaces, phase transitions, and multiphase flow.* Rev. Mod. Phys., vol. 66, no. 4, pages 1417–1479, 1994. 49

[Schweitzer 2003] F. Schweitzer. Brownian agents and active particles. Springer, Berlin, 2003. 64

[Shenderov 1997] A. D. Shenderov and M. P. Sheetz. *Inversely correlated cycles in speed and turning in an amoeba: An oscillatory model of cell locomotion.* Bioph. J., vol. 72, page 23822389, 1997. 22

[Sherratt 1992] J. A. Sherratt and M. A. Nowak. *Oncogenes, anti-oncogenes and the immune response to cancer: a mathematical model.* Proc. Roy. Soc. Lond. B, vol. 248, pages 261–271, 1992. 4

[Sherratt 2001] J. A. Sherratt and M. A. J. Chaplain. *A new mathematical model for avascular tumor growth.* J. Math. Biol., vol. 43, pages 291–312, 2001. 4

[Shlesinger 2003] M. F. Shlesinger. Supra-diffusion. in processes with longrange correlations. Berlin, Germany, Springer, 2003. 35

[Sisterson 2002] M. S. Sisterson and A. L. Averill. *Costs and benefits of food foraging for a braconid parasitoid.* J. Ins. Behav., vol. 15, no. 4, pages 571–588, 2002. 119

[Smallbone 2007] K. Smallbone, R. Gatenby, R. Gillies, P. Maini and D. Gavaghan. *Metabolic changes during carcinogenesis: Potential impact on invasiveness.* J. Theor. Biol., vol. 244, pages 703–713, 2007. 5

[Smoluchowski 1906] M. Smoluchowski. *Zur kinetischen Theorie der Brownschen Molekularbewegung und der Suspensionen.* Ann. Phys., vol. 21, pages 756–780, 1906. 23

[Spencer 2004] S. L. Spencer, M. J. Berryman, J. A. Garcia and D. Abbott. *An ordinary differential equation model for the multistep transformation to cancer.* J. Theor. Biol., vol. 231, no. 4, pages 515 – 524, 2004. 107, 108

[Stein 2007] A. M. Stein, T. Demuth, D. Mobley, M. Berens and L. K. Sander. *A Mathematical Model of Glioblastoma Tumor Spheroid Invasion in a Three-Dimensional In Vitro Experiment.* Biophys. J., vol. 92, no. 1, pages 356 – 365, 2007. 108

[Sullivan 2007] R. Sullivan and C. H. Graham. *Hypoxia-driven selection of the metastatic phenotype.* Canc. Metast. Rev., vol. 26, no. 2, pages 319–31, 2007. 118

[Swanson 2002] K. R. Swanson, E. C. Alvord Jr. and J. D. Murray. *Virtual brain tumors (gliomas) enhance the reality of medical imaging and highlights inadequacies of current therapy.* British Journal of Cancer, vol. 86, pages 14–18, 2002. 4, 5, 64, 122, 126

[Tektonidis 2008] M. Tektonidis. *Parameter optimization in a cellular automaton model of cancer growth based on medical mri data.* Master's thesis, TU Berlin, 2008. 121, 126, 131

[Tektonidis 2009] M. Tektonidis, H. Hatzikirou, A. Chauviere and A. Deutsch. *Unravelling the essential mechanisms of glioma tumor.* 2009. in preparation. 144

[Tucci 2005] K. Tucci and R. Kapral. *Mesoscopic multiparticle collision model for reaction-diffusion fronts.* J. Phys. Chem., vol. 109, page 21300, 2005. 94

[Turing 1952] A. M. Turing. *The chemical basis of morphogenesis.* Phil. Trans. R. Soc. Lond. B, vol. 237, page 3772, 1952. 61

[Turner 2002] S. Turner and J. A. Sherratt. *Intercellular adhesion and cancer invasion: A discrete simulation using the extended Potts model.* J. Theor. Biol., vol. 216, pages 85–100, 2002. 4

[Velikanov 1999] M. V. Velikanov and R. Kapral. *Fluctuation effects on quadratic autocatalysis fronts.* J. Chem. Phys., vol. 110, pages 109–115, 1999. 94

[Viswanathan 2008] G. M. Viswanathan, E. P. Raposo and M. G. E. da Luz. *Levy flights and superdiffusion in the context of biological encounters and random searches.* Phys. Lif. Rev., vol. 5, no. 3, pages 133 – 150, 2008. 119

[Werbowetski 2004] T. Werbowetski, R. Bjerkvig and R. F. Del Maestro. *Evidence for a secreted chemorepellent that directs glioma cell invasion.* Journal of Neurobiology, vol. 60, no. 1, pages 71–88, 2004. 118

[White 2005] S. M. White and K. A. J. White. *Relating coupled map lattices to integro-difference equations: dispersal-driven instabilities in coupled map lattices.* J. Theor. Biol., vol. 235, no. 4, pages 463 – 475, 2005. 33

[Wolf-Gladrow 2005] D. A. Wolf-Gladrow. *Lattice-gas cellular automata and lattice boltzmann models - an introduction.* Springer, 2005. 127, 180

[Wu 1994] X. Wu and R. Kapral. *Effects of molecular fluctuations on chemical oscillations and chaos.* J. Chem. Phys., vol. 100, pages 5936–5948, 1994. 46

[Wurzel 2005] M. Wurzel, C. Schaller, M. Simon and A. Deutsch. *Cancer cell invasion of normal brain tissue: Guided by Prepattern?* J. Theor. Med., vol. 6, no. 1, pages 21–31, 2005. 4

[Zama 2006] M. H. Zama, P. Matsudaira and D. A. Lauffenburger. *Understanding effects of matrix protease and matrix organization on directional persistence and translational speed in three-dimensional cell migration.* Ann. Biomed. Eng., vol. 35, no. 1, pages 91–100, 2006. 78

[Zwanzig 2001] R. Zwanzig. *Nonequilibrium statistical mechanics.* Oxford University Press, 2001. 142

Lattice-gas cellular automata models for the analysis of tumor invasion

Haralambos Hatzikirou

Abstract:
Cancer cells display characteristic traits acquired in a step-wise manner during carcinogenesis. Some of these traits are autonomous growth, induction of angiogenesis, invasion and metastasis. In this thesis, the focus is on one of the latest stages of tumor progression, *tumor invasion*. Tumor invasion emerges from the combined effect of tumor cell-cell and cell-microenvironment interactions, which can be studied with the help of mathematical analysis. Cellular automata (CA) can be viewed as simple models of self-organizing complex systems in which collective behavior can emerge out of an ensemble of many interacting "simple" components. In particular, we focus on an important class of CA, the so-called lattice-gas cellular automata (LGCA). In contrast to traditional CA, LGCA provide a straightforward and intuitive implementation of particle transport and interactions. Additionally, the structure of LGCA facilitates the mathematical analysis of their behavior. Here, the principal tools of mathematical analysis of LGCA are the mean-field approximation and the corresponding Lattice Boltzmann equation.

The main objective of this thesis is to investigate important aspects of tumor invasion, under the microscope of mathematical modeling and analysis:

Impact of the tumor environment: We introduce a LGCA as a microscopic model of tumor cell migration together with a mathematical description of different tumor environments. We study the impact of the various tumor environments (such as extracellular matrix) on tumor cell migration by estimating the tumor cell dispersion speed for a given environment.

Effect of tumor cell proliferation and migration: We study the effect of tumor cell proliferation and migration on the tumor's invasive behavior by developing a simplified LGCA model of tumor growth. In particular, we derive the corresponding macroscopic dynamics and we calculate the tumor's invasion speed in terms of tumor cell proliferation and migration rates. Moreover, we calculate the width of the invasive zone, where the majority of mitotic activity is concentrated, and it is found to be proportional to the invasion speed.

Mechanisms of tumor invasion emergence: We investigate the mechanisms for the emergence of tumor invasion in the course of cancer progression. We conclude that the response of a microscopic intracellular mechanism (migration/proliferation dichotomy) to oxygen shortage, i.e. hypoxia, maybe responsible for the transition from a benign (proliferative) to a malignant (invasive) tumor.

Computing *in vivo* tumor invasion: Finally, we propose an evolutionary algorithm that estimates the parameters of a tumor growth LGCA model based on time-series of patient medical data (in particular Magnetic Resonance and Diffusion Tensor Imaging data). These parameters may allow to reproduce clinically relevant tumor growth scenarios for a specific patient, providing a prediction of the tumor growth at a later time stage.

Keywords: Tumor invasion; mathematical modeling; lattice-gas cellular automata; Mean-field approximation; Lattice Boltzmann model

Zelluläre Gitter-Gas Automaten Modelle für die Analyse von Tumorinvasion
Haralambos Hatzikirou

Zusammenfaßung:
Krebszellen zeigen charakteristische Merkmale, die sie in einem schrittweisen Vorgang während der Karzinogenese erworben haben. Einige dieser Merkmale sind autonomes Wachstum, die Induktion von Angiogenese, Invasion und Metastasis. Der Schwerpunkt dieser Arbeit, liegt auf der *Tumorinvasion*, einer der letzten Phasen der Tumorprogression. Die Tumorinvasion ensteht aus der kombinierten Wirkung von den Wechselwirkungen Tumorzelle-Zelle und Zelle-Mikroumgebung, die mit die Hilfe von mathematischer Analyse untersucht werden können. Zelluläre Automaten (CA) können als einfache Modelle von selbst-organisierenden komplexen Systemen betrachtet werden, in denen kollektives Verhalten aus einer Kombination von vielen interagierenden "einfachen" Komponenten entstehen kann. Insbesondere konzentrieren wir uns auf eine wichtige CA-Klasse, die sogenannten Zelluläre Gitter-Gas Automaten (LGCA). Im Gegensatz zu traditionellen CA, bieten LGCA eine einfache und intuitive Umsetzung der Teilchen und Wechselwirkungen. Zusätzlich erleichtert die Struktur der LGCA die mathematische Analyse ihres Verhaltens. Die wichtigsten Werkzeuge der mathematischen Analyse der LGCA sind hier die Mean-field Approximation und die entsprechende Lattice - Boltzmann - Gleichung.

Das wichtigste Ziel dieser Arbeit ist es, wichtige Aspekte der Tumorinvasion, unter dem Mikroskop der mathematischen Modellierung und Analyse zu erforschen:

Auswirkungen des Tumorumgebung: Wir stellen einen LGCA als mikroskopisches Modell der Tumorzellen-Migration in Verbindung mit einer mathematischen Beschreibung der verschiedenen Tumorumgebungen vor. Wir untersuchen die Auswirkungen der verschiedenen Tumorumgebungen (z. B. extrazellulären Matrix) auf die Migration von Tumorzellen dürch Schätzung der Tumorzellen-Dispersionsgeschwindigkeit in einem gegebenen Umfeld.

Wirkung von Tumor-Zellenproliferation und Migration: Wir untersuchen die Wirkung von Tumorzellenproliferation und Migration auf das invasive Verhalten der Tumorzellen durch die Entwicklung eines vereinfachten LGCA Tumorwachstumsmodells. Wir leiten die entsprechende makroskopische Dynamik und berechnen die Tumorinvasionsgeschwindigkeit im Hinblick auf die Tumorzellenproliferation- und Migrationswerte. Darüber hinaus berechnen wir die Breite der invasiven Zone, wo die Mehrheit der mitotischer Aktivität konzentriert ist, und es wird festgestellt, dass diese proportional zu den Invasionsgeschwindigkeit ist.

Mechanismen der Tumorinvasion Entstehung: Wir untersuchen Mechanismen, die für die Entstehung von Tumorinvasion im Verlauf des Krebs zuständig sind. Wir kommen zu dem Schluss, dass die Reaktion eines mikroskopischen intrazellulären Mechanismus (Migration/Proliferation Dichotomie) zu Sauerstoffmangel, d.h. Hypoxie, möglicheweise für den Übergang von einem gutartigen (proliferative) zu einer bösartigen (invasive) Tumor verantwortlich ist.

Berechnung der *in-vivo* Tumorinvasion: Schließlich schlagen wir einen evolutionären Algorithmus vor, der die Parameter eines LGCA Modells von Tumorwachstum auf der Grundlage von medizinischen Daten des Patienten für mehrere Zeitpunkte (insbesondere die Magnet-Resonanz-und Diffusion Tensor Imaging Daten) ermöglicht. Diese Parameter erlauben Szenarien für einen klinisch relevanten Tumorwachstum für einen bestimmten Patienten zu reproduzieren, die eine Vorhersage des Tumorwachstums zu einem späteren Zeitpunkt möglich machen.

Keywords: Tumorinvasion; mathematische Modellierung; Zelluläre Gitter-Gas Automaten; Mean-field Angleichung; Lattice Boltzmann-Modell

Affirmation

I herewith declare that I have produced this paper without the prohibited assistance of third parties and without making use of aids other than those specified; notions taken over directly or indirectly from other sources have been identified as such. This paper has not previously been presented in identical or similar form to any other German or foreign examination board.

The thesis work was conducted from 07/2004 to 03/2009 under the supervision of Prof.Dr. Andreas Deutsch at ZIH, Department of Informatics, TU Dresden.

I declare that I have not undertaken any previous unsuccessful doctorate proceedings.

I declare that I recognize the doctorate regulations of the Fakultät für Mathematik und Naturwissenschaften of the Technische Universität Dresden.

Versicherung

Hiermit versichere ich, dass ich die vorliegende Arbeit ohne unzulässige Hilfe Dritter und ohne Benutzung anderer als der angegebenen Hilfsmittel angefertigt habe; die aus fremden Quellen direkt oder indirekt übernommenen Gedanken sind als solche kenntlich gemacht. Die Arbeit wurde bisher weder im Inland noch im Ausland in gleicher oder ähnlicher Form einer anderen Prüfungsbehörde vorgelegt.

Die Dissertation wurde von Andreas Deutsch at ZIH, Fakultät für Informatik, TU Dresden betreut und im Zeitraum vom 07/2004 bis 03/2009 verfasst.

Meine Person betreffend erkläre ich hiermit, dass keine früheren erfolglosen Promotionsverfahren stattgefunden haben.

Ich erkenne die Promotionsordnung der Fakultät für Mathematik und Naturwissenschaften, Technische Universität Dresden an.

Dresden, den 25 März 2009

Abbreviations

Abbreviation	Explanation
B.C.	Boundary Conditions
CA	Cellular Automaton
CML	Coupled Map Lattice
DTI	Diffusion Tensor Imaging
EA	Evolutionary Algorithm
FK	Fisher-Kolmogorov
GBM	Gliomblastoma Multiforme
LGCA	Lattice-Gas Cellular Automata
LBE	Lattice Boltzmann Equation
MF	Mean-Field
MRI	Magnetic Resonance Imaging
ODE	Ordinary Differential Equation
PDE	Partial Differential Equation

List of Symbols

Symbol	Explanation	Page
$\mathcal{L} \subset \mathbb{Z}^d$	d-dimensional regular lattice	12
$\mathcal{E}$	discrete state space	12
$\mathcal{R} : \mathcal{E}^{|\mathcal{N}|} \to \mathcal{E}$	local cellular automaton rule	12
$|\cdot|$	cardinality of a set	14
L_i, $i = 1, ..., d$	length of the lattice along i^{th} dimension	14
$(\mathbf{r}, k) \in \mathbb{N}^2$	discrete spationtemporal variables	14
$\mathbf{c_i} \in \mathbb{R}^d$, $i = 1, ..., b$	velocity channel vector	14
$\beta \in \mathbb{N}_0 = \mathbb{N} \cup \{0\}$	number of rest channels	14
b	coordination number	14
$\tilde{b} = b + \beta$	total number of channels	14
$\boldsymbol{\eta}(\mathbf{r}) \in \{0,1\}^{\tilde{b}}$	node configuration	14
$\eta_i(\mathbf{r}) \in \{0,1\}, i = 1, \ldots, \tilde{b}$	occupation number	14
$n(\mathbf{r},k) \in \{0, ..., \tilde{b}\}$	node density	14
$\boldsymbol{\eta}(k) := \{\boldsymbol{\eta}(\mathbf{r}, k)\}_{\mathbf{r} \in \mathcal{L}}$	global lattice configuration	14
$\mathcal{N}_b(\mathbf{r})$	set of b nearest neighborhood nodes	15
$\sigma \in \{0, ..., \varsigma\}$	σ species of a models with ς species	15
$\mathcal{R}^C$	interaction rule of a LGCA	15
$\mathbb{P}(\cdot)$	probability measure	15
$m \in \mathbb{N}$	single particle speed	15
$\tau \in \mathbb{N}$	automaton's time step	15
P	propagation operator	15
C_i, $i = 1, ..., b$	change in the occupation numbers	16
$\langle ... \rangle$	ensemble average	18
$|\Gamma| = 2^{\tilde{b}|\mathcal{L}|}$	discrete phase space	18
$\delta(x_0) = \delta(x, x_0) = \delta_{xx_0}$	delta function	18
$f_i(\mathbf{r}, k)$	single particle distribution	19
$\langle X^2 \rangle$	single particle mean square displacement	24
$\langle x^2 \rangle$	ensemble mean square displacement	25
O	reorientation operator	25
$\boldsymbol{\Omega}$	transition matrix of LBE	27
$\rho(\mathbf{r}, k) \in [0, \tilde{b}]$	mean node density	27
$(\mathbf{x}, t) \in \mathbb{R}^2 \times \mathbb{R}_+$	continuous spatiotemporal variables	31
$\varepsilon \ll 1$	scaling parameter	31
Γ	Boltzmann propagator	33
$\mathbf{q} = (q_i)$, $i = 1, ..., d$	discrete wave number	33
$\langle \cdot, \cdot \rangle$	inner product	34
R	Cell reaction (growth) operator	45
$\mathbf{J}(\boldsymbol{\eta}(\mathbf{r}, k))$	node flux	55
$\Theta(\cdot)$	Heaviside function	86

List of Figures

1.1 Hanahan and Weinberg have identified six possible types of cancer cell phenotypes: unlimited proliferative potential, environmental independence for growth, evasion of apoptosis, angiogenesis, invasion and metastasis (Reprinted from [Hanahan 2000], with permission from the authors.) 2

2.1 The sketch visualizes the hierarchy of relevant scales for the LGCA models introduced in this thesis (see text for details). 13

2.2 Node configuration: channels of node **r** in a two-dimensional square lattice ($b = 4$) with one rest channel ($\beta = 1$). Gray dots denote the presence of a particle in the respective channel. 15

2.3 Example of a possible interaction of particles at a two-dimensional square lattice node **r**; filled dots denote the presence of a particle in the respective channel. No confusion should arise by the arrows indicating channel directions. 16

2.4 Propagation in a two-dimensional square lattice with speed $m = 1$; lattice configurations before and after the propagation step; filled dots denote the presence of a particle in the respective channel. 17

3.1 Reorientation rule of random motion. The first column corresponds to the number of cells on a node $n(\mathbf{r}, k)$ at a time km, with capacity $\tilde{b} = 4$. The right column indicates all the possible cell configurations on node and the transition probability of obtaining a certain configuration (3.10). 26

3.2 The left figure presents the trajectory of a randomly moving cell. The right figure shows the time evolution of the corresponding mean square displacement. The lattice consist of nodes with $\tilde{b} = 6$. The expected diffusion coefficient of an individual cell is $D^* = \Delta \langle X_k^2 \rangle / b = 0.67/4 \simeq 1/6$, where $b = 4$ the dimension of the system. 29

3.3 Temporal evolution of a diffusive cell cluster at times $t = 0, 100, 200$. The initial configuration represents a fully occupied circle of nodes. Colors indicate the node density (see colorbar). 31

3.4 Illustration of the analogy between the evolution of a system on a lattice with periodic B.C. on axis-L_2 and the evolution of a system on a cylindrical surface ("tube"). For more details see the text (subsec. 3.3.2.5). 36

3.5 Typical simulation on a square lattice with $\tilde{b} = 6$ channels with periodic boundary condition along the L_2-axis. The colors denote the node density. In the middle the red stripe represents the initial conditions of the simulation, which is a fully occupied lattice column of cells. 37

3.6 The left figure represents the averaged quantity $n(r_x, k)$ of fig. 3.5 (where $\tilde{b} = 6$). On the right, we observe the time evolution of the mean square displacement. The diffusion coefficient of the simulations, which is the slope of the line, coincides with the theoretically calculated value $D = m^2/\tilde{b}\tau = 1/6 \simeq 0.17$. 38

4.1 Typical simulations of the spatiotemporal evolution of the LGCA growth process starting from an initial fully occupied cluster of nodes in the center of the lattice. The three figures show snapshots of the same simulation at different times. The colors encode the node density. 56

4.2 Evolution of the per capita growth γ as a function of the total population density $\bar{\rho}$. Observe that the γ calculated from the LGCA simulations, decreases rapidly for increasing population densities. The behavior of γ can be fitted by a curve $A\bar{\rho}^{-1/2}$, as it is calculated in our spatio-temporal MF analysis (4.69). Additionally, we observe that the temporal MF (4.66) completely fails to follow the actual LGCA dynamics.The log-log plot (right figure) allows for a better distinction between the fit and the simulation curves, especially for low $\bar{\rho}$. 57

5.1 An example of a vector field (tensor field of rank 1). The vectors (e.g. integrin receptor density gradients) show the direction and the strength of the environmental drive. 66

5.2 An example of a tensor field (tensor field of rank 2). We represent the local information of the tensor as ellipsoids. The ellipsoids can encode e.g. the degree of alignment of a fibrillar tissue. The colors are denoting the orientation of the ellipsoids. 67

5.3 Time evolution of a cell population under the effect of a field $\mathbf{E}=(1,0)$. One can observe that the environmental drive moves all the cells of the cluster into the direction of the vector field. The blue color stands for low, the yellow for intermediate and red for high densities. 69

5.4 The figure shows the evolution of the cell population under the influence of different fields (100 time steps). Increasing the strength of the field, we observe that the cell cluster is moving faster in the direction of the field. This behavior is characteristic of a haptotactically moving cell population. The initial condition is a small cluster of cells in the center of the lattice. Colors denote different node densities (as in fig. 5.3). 71

5.5 Time evolution of a cell population under the effect of a tensor field with principal eigenvector (principal orientation axis) $\mathbf{E}=(2,2)$. We observe cell alignment along the orientation of the axis defined by E, as time evolves. Moreover, the initial rectangular shape of the cell cluster is transformed into an ellipsoidal pattern with principal axis along the field $\mathbf{E}$. Colors denote the node density (as in fig. 5.3). . . . 72

5.6 In this graph, we show the evolution of the pattern for four different tensor fields (100 time steps). We observe the elongation of the ellipsoidal cell cluster when the strength is increased. Above each figure the principal eigenvector of the tensor field is denoted. The initial conditions is always a small cluster of cells in the center of the lattice. The colors denote the density per node (as in fig. 5.3). 73

5.7 We show the brain's fiber track effect on glioma growth. We use a LGCA of a proliferating cancer cell population (for definition see [Hatzikirou 2008a]) moving in a tensor field provided by clinical DTI data, representing the brain's fiber tracks. **Top**: the left figure is a simulation without any environmental information (only diffusion). In the top right figure the effect of the fiber tracks in the brain on the evolution of the glioma growth is obvious. **Bottom**: The two figures display magnifications of the tumor area in the simulations above. This is an example of how environmental heterogeneity affects cell migration (where in this case tumor cell migration). 75

5.8 This figure shows the variation of the normalized measure of the total lattice flux $|\mathbf{J}|$ against the field intensity $|\mathbf{E}|$,where $\mathbf{E} = (e_1, e_2)$. We compare the simulated values with the theoretical calculations (for the linear and non-linear theory). We observe that the linear theory predicts the flux strength for low field intensities. Using the full distribution, the theoretical flux is close to the simulated values also for larger field strengths. 77

5.9 The figure shows the variation of the X-Y flux difference against the anisotropy strength (according to Model II). We compare the simulated values with the linear theory and observe a good agreement for low field strength. The range of agreement, in the linear theory, is larger than in the case of model I. 79

6.1 **Left**: Typical spatio-temporal pattern formation of *in vitro* tumors (reprinted with permission from Folkmann et al. [Folkman 1973]). One observes clearly the presence of a necrotic core and an outer rim of proliferative tumor cells. **Right**: A LGCA simulation exhibits a similar structure. In the simulation, tumor cells are depicted in grey, necrotic entities in white, and empty nodes in black. The comparison of the two figures is phenomenological, at the level of pattern formation, and not quantitative. 87

6.2 **Left**: Two corresponding nodes at position **r**, one from the tumor and the other from the necrotic lattice. The grey stripe denotes one chosen couple of channels. **Right**: Spatio-temporal pattern formation in the LGCA model. An invading two-dimensional tumor wavefront for $r_M = 0.2$ and $r_N = 0.7$. Tumor cells are depicted in grey, necrotic entities in white, and empty nodes in black. 88

6.3 **Left**: Snapshot of the average concentration profile along the L_1-axis, i.e. $n_x(k) = n(r_x, k) = \frac{1}{|L_2|} \sum_{r_y \in |L_2|} n(\mathbf{r}, k)$. **Right**: Linear growth of the tumor front distance from its initial position, denoted as front position. The slope of the line defines the speed of the tumor invasion. 89

6.4 The second steady state solution of the tumor cells $\bar{f}_C = g(r_M, r_N)$ for different values of mitotic and necrotic probabilities. The mitotic and the necrotic thresholds are $\theta_M = 4$ and $\theta_N = 6$, respectively. . . 91

6.5 A sketch of the wavefront as shown in fig. 6.3 (left). We distinguish three regimes: (i) $x \in [x_\delta, x_0]$, where $0 < \rho(x) < \delta$: the region represents a highly fluctuating zone, where the cells perform a random walk with almost no proliferation, (ii) $x \in [x_C, x_\delta]$, where $\delta < \rho(x) < C$: this region is a result of non-linear proliferation and cell diffusion and (iii) $x \in [0, x_C]$, where $\rho(x) \simeq C$: this regime represents the bulk of the front (saturated lattice) where no significant changes are observed. 95

6.6 Comparison of the calculated front speed for the naive and the cut-off MF, i.e. v_n and v_c respectively, against simulations. We observe that the cut-off MF predicts closely the front speed calculated from the simulations for $K \simeq 1.7$. 98

6.7 Numerically the front width is estimated by fitting a straight line, tangential to the inflection point $\rho_C(x^*) = \bar{b}g/2$ and the front width W is approximated as the inverse slope of the fitted line. In this example, the simulation time is 1500 steps, the mitotic rate $r_M = 0.03$, necrotic rate $r_N = 0.7$, $|L_2| = 10$ and $g \simeq 0.38$ (it can be estimated from fig. 6.4). The black, dotted lines are indicating the methodology followed for the calculation of W. 99

7.1 This flow diagram shows how the oxygen concentration influences the phenotype of a tumor cell. The "Go or Grow" mechanism respond to low oxygen supply with a motile phenotype. On the contrary, high oxygen level contributes to the occurrence of the proliferative phenotype. 104

7.2 T1-weighted contrast-enhanced coronal MRI of recurrent glioblastoma. **Left**: T1-weighted contrast-enhanced coronal MRI section showing the resection cavity in the right parietal lobe – 3 months postoperatively. There is no evidence of tumor recurrence. **Middle**: corresponding MRI of the same patient, 6 months postoperatively, clearly showing tumor recurrence (=hyperintense or white mass). **Right**: control MRI, 9 months postoperatively, showing tumor extension beyond the previous resection cavity along the cerebral white matter. 105

7.3 Sketch of pre- and postoperative of a glioblastoma tumor. In the preoperative state a GBM tumor constitutes of an inner core of proliferative cells and an outer ring of invasive tumor cells (for simplification reasons the necrotic core is neglected). After a gross total resection, the main part of the tumor disappears but some cells of the invasive zone stay intact. These cells are assumed to be responsible for the tumor recurrence. 106

7.4 Diagrammatical representation of mutation-driven phenotypic changes (from the proliferative to the invasive phenotype). Cells that belong to the proliferative population (ρ_p) undergo mitosis and the newborn cells ($r_p \rho_p$) are subjected to mutation events. The mutation of the appropriate combination of genes leads to the change of phenotype in $mr_p \rho_p$ cells. 107

7.5 This flow-chart represents how the mitosis/apoptosis operator is modeled. The algorithm is applied for each node of the automaton. Initially, the algorithm checks if the node is empty. If not, cells can undergo mitosis or apoptosis depending on the probabilities (7.5) and (7.6), respectively. In the case of apoptosis the node looses a cell. On the other hand, mitosis is conditioned by the existence of at least one free channel. If the node is not full then the newborn cell is added to the node. 110

7.6 Typical pattern formation after 1000 time steps. For maximum occupancy $C = \tilde{b}$ and for a fixed proliferation rate $r_m = 0.05$, we vary the number of rest channels $\beta = 2, 4, 6$. We observe that, starting from an initial localized occupation, the motile populations (low β) expand faster than the proliferative ones. The colors encode the node density. 112

7.7 Infiltration radius against phenotype. For maximum occupancy $C = \tilde{b}$ and for a fixed base proliferation rate $r_m = 0.05$, we show how the invasive radius varies by increasing the number of rest channels β, i.e. by making the tumor cells more proliferative. We observe a linear decrease of the infiltration radius as β increases. 113

7.8   Typical simulation on a "tubular" lattice, i.e. with periodic boundary condition along the y-axis. The colors denote the node density. In the mid-section of the figure, the white part denotes nodes with maximum density. 114

7.9   On the x-axis we vary the parameter β, which characterizes the tumor cell phenotype, ranging from motile populations (β small) to proliferative ones (β large). Each of the curves represents an iso-nutrient, i.e. the behavior of the population under the same oxygen availability. We observe that each iso-nutrient curve has a maximum point, which corresponds to the best fitted phenotype (β) in this specific environmental setting. 115

8.1   Illustration of the algorithm. The gray boxes represent the four main modules and the yellow boxes the data. The segmentation module operates only once for each MR image. The evolutionary algorithm, the lattice-gas cellular automaton simulation and the shape analysis are involved in the main loop (bold arrows) and exchange data. The algorithm terminates when the EA delivers an acceptable parameter set. 125

8.2   Example of a segmentation. 126

8.3   Channels of a node in the hexagonal lattice of the *two-speed model* [Wolf-Gladrow 2005]. 127

8.4   Hexagonal coordinate system described by the unit vectors **u** and **v**. The angle between **u** and **v** is 60°, while the corresponding angle in the orthogonal coordinate system between **x** and **y** is 90°. 130

8.5   Shape analysis of four images. 131

8.6   Data of experiment: **(a)** Tumor of *time point A* mapped (red) on the vector field of the DTI data of a brain. The initial tumor has been placed on fiber tracks. **(b)** Tumor of *time point B* mapped on the vector field; note the simulated tumor growth along the fiber tracks. 135

8.7   Fitness function applied to some simulation results; the red color indicates the simulated tumor growth and the white line indicates the boundary of the tumor shape (image of *time point B*) **(a)** A simulation using the optimal parameter set achieves a high fitness value. **(b)** A similar parameter set to the optimal one, achieves a high fitness value. **(c-d)** Parameter sets that differ significantly from the optimal one achieve lower fitness values. 136

8.8   Experiment 2: Evolution of the population fitness. 136

VDM Verlagsservicegesellschaft mbH

Die VDM Verlagsservicegesellschaft sucht für wissenschaftliche Verlage abgeschlossene und herausragende

Dissertationen, Habilitationen, Diplomarbeiten, Master Theses, Magisterarbeiten usw.

für die kostenlose Publikation als Fachbuch.

Sie verfügen über eine Arbeit, die hohen inhaltlichen und formalen Ansprüchen genügt, und haben Interesse an einer honorarvergüteten Publikation?

Dann senden Sie bitte erste Informationen über sich und Ihre Arbeit per Email an *info@vdm-vsg.de*.

Sie erhalten kurzfristig unser Feedback!

VDM Verlagsservicegesellschaft mbH
Dudweiler Landstr. 99   Telefon +49 681 3720 174
D - 66123 Saarbrücken   Fax   +49 681 3720 1749

www.vdm-vsg.de

Die VDM Verlagsservicegesellschaft mbH vertritt

Printed by Books on Demand GmbH, Norderstedt / Germany